Quantum Entanglement

and

The Loss of Reality

Thomas V Marcella

Thomas V Marcella is Professor Emeritus in Physics and Applied Physics at the University of Massachusetts Lowell.

Copyright © 2018 by Thomas V Marcella

All rights reserved.

ISBN-10:198180742X

ISBN-13:978-1981807420

COVER PHOTOS
Albert Einstein lecturing in Vienna, 1921.
Photographer: Ferdinand Schmutzer.
Wikimedia Commons/Public Domain.

Niels Bohr as a young man.
Exact date of photo unknown.
Wikimedia Commons/Public Domain.

FOR MY FAMILY

IN MEMORIAM

Leon Beghian   *University of Massachusetts Lowell*

Robert Carovillano   *Boston College*

ACKNOWLEDGEMENT

I wish to thank Malcolm "Mac" Knapp for his critical reading of an earlier version of this manuscript. His insightful comments and suggestions are gratefully acknowledged.

CONTENTS

Preface

1   Introduction
- 1.1   General   1
- 1.2   Classical physics vs quantum physics   2
- 1.3   Schrödinger's cat experiment   9
- 1.4   The classical description of the Schrödinger cat experiment   11
- 1.5   The quantum description of the Schrödinger cat experiment   15

2   The Classical Universe
The world view of Albert Einstein
- 2.1   General   21
- 2.2   The total solar eclipse of 21 August 2017   22
- 2.3   Classical physics is about reality   25
- 2.4   Objective reality   26
- 2.5   Real things exist in space-time   28
- 2.6   The locality principle   29
- 2.7   The principle of separability   30
- 2.8   Causal determinism   32
- 2.9   The classical single-slit experiment   35

3 The Decline of Classical Physics
The essential quantum discontinuities make their appearance

3.1	Particles and waves in classical physics	40
3.2	Light is an electromagnetic wave	41
3.3	The first mention of 'quanta': Planck's law and blackbody radiation	41
3.4	Light consists of particles: The photoelectric effect	47
3.5	So, is light a wave or is it made up of particles?	51
3.6	Bohr's 1913 description of the hydrogen atom	53
3.7	Electron capture	58

4 The Quantum Universe
The world view of Niels Bohr

4.1	Complementarity vs classical realism	59
4.2	The principle of non-separability (entanglement)	62
4.3	A quantum experiment requires a result	68
4.4	Which preparation procedure? Delayed choice	70
4.5	The use of classical concepts	71
4.6	The use of ordinary language	73
4.7	The Stern-Gerlach Experiment	74
4.8	Spin vectors and a common trig identity	80

5 The postulates of quantum mechanics
A set of instructions for doing quantum calculations

5.1	*General*	87
5.2	*The postulates of quantum mechanics*	89
5.3	*Quantum uncertainty*	98
5.4	*The Heisenberg uncertainty principle*	103
5.5	*de Broglie's hypothesis*	108
5.6	*The single slit scattering experiment: particle diffraction*	110

6 Quantum interference
The role of entanglement in particle interference

6.1	*General*	115
6.2	*Particle interference*	116
6.3	*The generic interference experiment with particles*	118
6.4	*The superposition of probability amplitudes*	129
6.5	*Constructive and destructive interference*	131
6.6	*Complementarity and quantum interference*	131
6.7	*The Mach-Zehnder interferometer*	134
6.8	*The double slit experiment*	138
6.9	*Interference with a modified Stern-Gerlach experiment*	142

7 Einstein, Podolsky, and Rosen
An attempt to show that quantum mechanics is incomplete

7.1	*General*	147
7.2	*Einstein's boxes*	149
7.3	*Overview of the EPR experiment*	156
7.4	*The EPR experiment according to EPR*	159
7.5	*The EPR experiment according to Bohr*	165
7.6	*The Bohm version of the EPR experiment*	170

8   Bell's theorem
The second revolution begins
8.1   Introduction to Bell's theorem   173
*8.2   A Bell-like experiment - comparing
        true-false test results*   176
8.3   Bell's experimental test for local realism   181
8.4   The quantum calculation with entanglement 192
8.5   The empirical evidence   196

Further Readings   *201*

PREFACE

Fresh out of the army in September of 1957 I entered the Lowell Technological Institute planning on becoming an electrical engineer. In the army I had been a radar repairman. I enjoyed it. Electronics was fun. But I never did get to study electrical engineering in college.

On 4 October 1957, the Soviet Union launched Sputnik I. The space race was on! Physics was in the news. After a brief discussion with the head of the physics department, I changed my major to physics. I have Sputnik I to thank for my career as a physics professor.

As undergraduates, budding physicists traditionally learn classical theory first while developing the required mathematical skills. What you learn today depends on what you did yesterday. Classical physics is neat and logical and consistent. We have an intuitive feeling about it. In principle, there is nothing unusual going on in the classical universe. Slowly, but surely, you begin to develop some confidence in what you are doing.

But that all came to an end with the first course in quantum mechanics, which is entirely different. In fact, the classical mindset acquired in the first few years of study becomes an obstacle to understanding the new theory.

Sometime around 1960, we met Donald Cunningham, a recent Physics PhD who came to Lowell to help us form a student section of the American Institute of Physics.

Over lunch, he confided to us that he and his fellow graduate students considered their quantum mechanics class to be "the mystery hour". This was an eye-opener for us. We had imagined that we were struggling in isolation. Now we were being told that others, who eventually succeeded in learning physics, also struggled along the way. We were not alone!

It gave me further assurance when Richard Feynman said, "I think I can safely say that no one understands quantum mechanics." Apparently it was acceptable to not understand.

Our introductory quantum mechanics course was all about doing the necessary calculations. The troubling aspects were ignored. Einstein, Podolsky, and Rosen were never mentioned. We knew nothing of any Bohr-Einstein debate. Entanglement, as far as we knew, had nothing to do with quantum physics. And, Bell had not yet done his seminal work.

Years later, I would continue this tradition by teaching quantum mechanics in the same way. Now, this book gives the teacher in me a second chance to discuss some of those long ignored features that still cause so much consternation. Most of what you will find here was not taught when I was an undergraduate.

This book is intended to be a popular interest book in quantum mechanics that emphasizes the fundamental character of entanglement in quantum theory. And with entanglement comes all the other absurd aspects of quantum mechanics. I hope it will be of interest to professional physicists and physics students, as well as engineers, science teachers, and other intellectually curious laymen.

The real difficulty in teaching these things lies in the fact that the atomic world does not resemble the physical world of our everyday experience. The atomic domain is not just a tiny version of the macroscopic universe that is perfectly described by the classical laws of Newton and Maxwell, which describe the behavior of real things. And there is no doubt in our minds what is real and what is not. It is intuitively obvious.

We accept classical physics to be a true representation of the physical universe as it is. It would be much easier for us if atomic things were also a part of that physical reality. But they are not! If we try to scale down to the level of atoms and the like, what we call 'physical reality' loses its usual meaning. For example, it is now generally accepted by the physics community that electrons are not real in the same way that a baseball is. I am confident that classical physics correctly describes the behavior of baseballs in all situations. But the same cannot be said for electrons. Classical theory cannot explain what is happening to electrons in all circumstances.

Classical physics, the physics of real things, fails on the atomic level. The once powerful deterministic equations of Newton and Maxwell are of little help. Quantum things behave in a counterintuitive way unfamiliar to us and, when we try to explain the details of an atomic event, we often find ourselves talking nonsense. Many of the best minds in science have been unable to reach a consensus on the meaning of it all.

Obviously, a radical revision of scientific thought is required to adequately describe atomic events, and the new theory is nothing like its predecessor. A new approach to science replaces the old classical methods, and it requires a great deal of mental discipline to accept the alien concepts of the new theory. And that is a big part of the difficulty as we transition to the new way of thinking.

Also adding to the difficulty is our strong classical bias. We bring a ready-made set of classical ideas to the quantum discussion. We think we know how atoms are supposed to behave.

From its very beginning, quantum mechanics forced us to question the most fundamental beliefs of classical physics. First Planck, then Einstein and Bohr, suggested that our understanding of the physical world was seriously flawed. Controversy, it seems, has always been part of the quantum discussion.

Most perplexing was the strange concept to be named 'entanglement' by Schrödinger shortly after the 1935 paper of Einstein, Podolsky, and Rosen. It was EPR, as they came to be known, that introduced us to certain long distance correlations that have no reasonable explanation. The quantum description suggests that doing something here can instantaneously change something far away. Such non-local, voodoo-like, interactions have always been anathema to the physics community. Einstein called them "spooky action-at-a-distance".

But Bohr argued that under certain conditions widely separated quantum particles are actually non-separable and one particle can instantaneously 'influence' its far away partner. In quantum theory they are not widely separated as seen in the laboratory. Rather, they are a single quantum entity, connected in some mysterious way.

Bohr's 'influence' is Schrödinger's 'entanglement' and, for Bohr, entanglement is accompanied by such non-local interactions. According to Bohr, two or more entangled particles constitute a wholeness unlike anything in the real world. Once entangled, this wholeness persists even after the particles have been spatially separated over a vast distance. Einstein thought this was nonsense.

We had to wait until 1964 when Bell and his famous inequality would show us how to proceed in order to resolve the dispute.

We take the reader from classical physics through Bell's theorem in the context of the Bohr-Einstein debate over the meaning of reality in quantum mechanics. We present the classical approach of Einstein against the quantum mechanics of Bohr, pitting common sense against the counterintuitive nature of the new theory. In particular, we discuss the EPR experiment and Bell's theorem in detail. At the end of it all, we are forced to conclude, as did Bell, that there is no realistic way for a quantum event to happen.

There is no explanation of how a quantum experiment works. It just is! The results obtained in a quantum experiment occur randomly and without reason. Entanglement is the new norm, and with entanglement there is no reality as we know it.

We emphasize throughout the text that if we insist that quantum things are real, and then apply the laws of classical physics, our calculations do not agree with the experimental results. Classical physics simply does not work at the atomic level.

Although there is still some controversy over what Einstein and Bohr might have had in mind when discussing these things, we will attribute classical physics to Einstein and quantum theory to Bohr. Certainly, in the debate between them, Einstein defended the traditional classical view while Bohr was the proponent for the new quantum theory. Examples of this conflict are presented throughout the book.

To do this, we need to know what makes an event 'classical'. So, in Chapter 2 we elaborate on the main characteristics of the classical description of nature that are established in the laws of Newton, Maxwell, and Einstein.

What went wrong with classical theory is discussed in Chapter 3. Planck, Einstein, and Bohr were the instigators. Each introduced discontinuities into the discussion of atomic physics while solving different problems. Even here, at the beginning, some were bothered by the non-local nature of these discontinuous events. For the first time the well established classical description of light in terms of continuous waves was found to be wanting.

Planck introduced us to the 'elementary quantum of action' h, also known as Planck's constant, while making the outrageous claim that light exchanged energy with matter in discontinuous bursts of energy called 'quanta'.

Planck himself was uncomfortable with what he had done. He doubted the reality of his quanta. For him, his hypothesis was simply an ad hoc mathematical construct that led to the known answer he was looking for.

Entanglement and all the counterintuitive characteristics of quantum physics are consequences of the small, yet finite, size of Planck's constant. The discovery of the quantum changed physics forever.

The discussion in Chapter 4 is an introduction to quantum theory as I think Bohr would have done. We present the fundamental characteristics of quantum mechanics that distinguish it from classical physics. It is here that we present complementarity as Bohr's interpretation of quantum mechanics. We emphasize that the essential element of quantum theory is the quantum experiment, an entangled wholeness that has no classical analog.

But quantum mechanics is not entirely free of classical physics. Quantum theory still employs the concepts and terminology of its predecessor science, but the words used in the quantum description no longer have the same meanings as their classical counterparts. In Chapters 4 and 5 we caution the reader about the inherent confusion that results when we use the same word in both the classical and quantum contexts.

We affirm that the use of classical concepts and the use of ordinary language cannot adequately describe a quantum event. But there is no alternative.

Quantum mechanics has a mathematical structure considered to be rather abstract. It is! Yet, much of our understanding, as well as the vocabulary used, comes from the mathematical formalism. This presents a dilemma for anyone who writes about quantum mechanics for a general audience. Nevertheless, in Chapter 5, I have opted to present some math in the form of postulates that are used to do quantum mechanics. We do not expect the reader to do any calculations, but hopefully, this axiomatic approach will give my non-physicist readers a sense of quantum mechanics that otherwise would be denied them.

The last three chapters make up the core material of the book. We compare the predictions of classical physics to that of quantum mechanics for some specific experiments. We establish entanglement as the dominant feature of the quantum description.

In Chapter 6 we show how entanglement enables particle interference. Chapter 7 is a detailed analysis of the EPR experiment. In Chapter 8, we discuss Bell's failed attempt to introduce reality into quantum mechanics. We show that Bell's modified EPR experiment lends itself easily to both classical and quantum analysis. But, Bell concluded that quantum mechanics does not have a classical explanation. There is no underlying reality that can generate the same experimental results as quantum mechanics.

Most now believe that Bell and ensuing experiments have dashed any hope that atoms are real in the classical sense. We are forced to accept that atomic physics is indeed incompatible with an objective reality. Bell, like Einstein, had hoped that, even though atomic events are inherently statistical, they do have an underlying physical reality. They don't! There is now too much empirical evidence to think otherwise.

I will try to convince my readers that quantum mechanics is a stand alone science to be accepted as it is without trying to give it a classical interpretation. Unfortunately, this is extremely difficult to do in practice. We all know the frustration that comes when trying to describe quantum events.

When doing quantum experiments we can only be sure of results and probabilities. The ultimate question is, "Is this all there is to know about the quantum experiment?" Bohr would most likely answer, "Yes. If we know the results and the probability of occurrence for each result, then we know everything there is to know about that event. There is nothing else!"

"Not so," says Einstein. "Surely, there must be more to an experiment than just the results. Something has to be going on in the experiment before we get the result. Things don't just happen all by themselves. Obviously, quantum mechanics does not tell us the whole story."

Bell's theorem says they cannot both be correct. There can be no quantum mechanics that embraces the objective reality of classical physics. Nature has to choose one or the other.

Real world experiments have conclusively shown that nature does obey the rules of quantum mechanics. At its most fundamental level the world is probabilistic, not real as we imagine it to be.

When we are done, I hope my readers will have come to appreciate quantum mechanics, even if we don't understand it in a conventional sense, the way we understand classical physics.

"It is acceptable to not understand."

1

Introduction

Insanity: doing the same thing over and over again and expecting different results.

Albert Einstein

1.1 General

So, why should anyone want to write another book trying to explain quantum mechanics, when there are already many very good books in print? Obviously, I want this book to be different. How? Like Sgt. Joe Friday, I will try to focus on "just the facts". My intention is to describe some tell-tale experiments that disclose the quantum nature of atomic physics. If I am successful, then maybe you might come to accept this counterintuitive theory for what it is, and not try to force it to be something it is not.

I will try to do this by taking quantum mechanics at face value and by trying to avoid the many questions of interpretation, other than the complementarity principle of Bohr and the Born probability interpretation. Although the Born interpretation has near universal acceptance, there is still some controversy over what the Bohr interpretation really is.

If you have already looked at other books, you know that many people do spend a great deal of time talking about interpretation and trying to figure out what is really happening in a quantum event. I am not going to do that. I will leave that discussion to the philosophers among us. Instead, I will try to present what I consider to be an honest overview of Bohr's ideas. And I will try not to speculate about those things that have no basis in the theory and which cannot be verified experimentally.

I say 'try' because this is no easy task. Quantum physics is hard! There is much that makes it difficult to accept. To make matters worse, it is near impossible to describe with our ordinary language.

So if you are here expecting me to tell you how Schrödinger's cat can be alive and dead at the same time, then you are in the wrong place. I have no intention of doing that. In spite of what others have said, and in spite of what you may have read elsewhere, quantum mechanics does not suggest any such behavior. In fact, we shall see that the mathematical formalism specifically prohibits such a thing from happening, and no experiment has ever produced a cat that is alive and dead at the same time.

1.2 Classical Physics vs Quantum Physics

This is a book about particle physics in the atomic domain. But, we must accept that atomic particles are not ordinary. And that is the quandary in which we find ourselves. The world of atoms, electrons and photons is unlike anything we have ever encountered in our everyday experience.

There is no way that we can possibly imagine what it is really like. If we are to succeed in our endeavor, then we must abandon all hope of including such quantum things into our everyday description of nature. Bell tried and it didn't work! We must accept, as fact, that quantum particles do not obey the laws of Newton, Maxwell, and Einstein. Electrons, for example, are not everyday kind of charged things that just happen to be tiny.

Bohr separated the universe into two parts, the classical and the quantum. I hope to maintain that dichotomy between those two camps. As far as I know, he never did tell us where one ends and the other begins, but we shall not bother ourselves with that. Generally, we identify the classical world to be of macroscopic size, the size of things familiar to us. The classical world consists of the real things that make up our everyday experience, and those real things obey the deterministic laws of classical physics. On this level, the size of Planck's constant is negligible. It has no effect on classical physics.

Atoms, on the other hand, inhabit a microscopic domain, too small for us to imagine. The radius of a hydrogen atom is of the order of 10^{-10} meters, about a million times smaller than the width of a human hair. Other quantum particles are even smaller. An electron, for example, is thought to be a point particle, infinitesimally small, meaning it has no size at all! So, we are talking here of very small things, indeed. It is here that the non-zero value of Planck's constant becomes significant. These are the quantum things that obey the probabilistic laws of quantum mechanics. These are the things to be studied in this book.

Prior to 1900, the many successes of classical physics embraced all of the known universe, which is out there, all by itself, available for us to explore. It can be divided into its many parts. Each part is real with its own separate existence in space-time.

Such an external real existence does not depend on physicists, or their theories, or on their instruments of observation. After all, the universe was here before we were. It is the intention of classical physics to provide an understanding of this objective universe.

This belief in an objective reality is the paramount tenet of classical physics and Einstein believed there is no classical physics without it. For him, all of science, not only physics, depends on it.

All of the different parts of the classical universe are known to move smoothly through space under the influence of forces. Each part can be studied individually, and anything that happens, does so for a reason that is explained in the deterministic laws of Newton and Maxwell. Physicists of that era confidently concerned themselves with applying the physical laws to this observed physical reality.

But, unbeknown to most physicists, this world view of a smooth running mechanical universe was about to be shattered.

In the early 1900's, two grand theories were unleashed on a generally unsuspecting community of scientists. These were the special theory of relativity, conceived by Albert Einstein in 1905, and quantum theory, discovered by Max Planck in 1900 and developed into a working theory by Niels Bohr, Werner Heisenberg, Irwin Schrödinger and others during the mid 1920's. The mathematical formalism of quantum mechanics used to this day was pretty much completed by the end of 1927.

The special theory of relativity extends Newtonian physics into the realm of objects moving at speeds close to the speed of light and of events in which the equivalence of mass and energy is evident. Although quite radical in its own right, it did not change the philosophical mindset of the classical physicist. Einstein's relativistic universe is real. The classical belief in an objective reality remains intact in relativity theory. Although there was some resistance to it initially, it is safe to say that the physics community now accepts relativity theory without controversy.

But, quantum mechanics is something else entirely. It has been ninety years since its conception and the upheaval it caused is still with us. In spite of unparalleled success, we still do not know what it all means! We don't even know what it is that we are trying to describe. There are still many unanswered questions, in spite of the efforts of many of the best minds in science.

Atomic objects are not like the ordinary objects we are familiar with. Worse still, they exist only in conjunction with the experimental apparatus used to observe them. In Bohr's world, there are no atoms, only experiments done with atoms.

Quantum mechanics does not describe the behavior of individual things on the atomic scale in the same way that we describe macroscopic objects. In fact, the new theory denies any objective reality to the quantum universe. There is no quantum universe 'out there' for us to explore. An electron, for example, does not exist 'all by itself'. This causes great consternation and lack of agreement among many physicists, and philosophers alike. If you are uncomfortable with all this, you are in good company. Our protagonists, Einstein and Bohr struggled with these counterintuitive suggestions their entire lives.

Quantum mechanics is all about doing experiments. But it predicts only the possible results and the probability of obtaining each result when we make a measurement. That's all there are!

While classical physics concerns itself with 'what is really happening' to objects themselves, quantum mechanics describes only experiments and their observed results. And, in the quantum experiment, as strange as it seems, there is nothing happening before we make the actual measurement. Classical physics tells us how experiments work, while quantum mechanics does not.

We know that the probabilistic predictions of quantum mechanics are correct. The theory is far too successful to think otherwise. Much of modern technology would not exist if it weren't for knowledge obtained from quantum mechanics.

All of the known facts about atoms and the like are correctly predicted by the theory. It has yet to fail us. But, the theory has forced us to reject the most sacred beliefs of classical thought. What is considered to be 'common sense' in our everyday thinking is not only unreliable on the atomic level, but it must be rejected outright, lest we get dragged into a morass of contradiction, paradox, and general confusion. Our intuition fails us in the quantum domain.

Classical physics is about reality. Quantum physics is not.

So, how do we avoid the pitfalls that beleaguer those trying to understand quantum mechanics? Bohr suggests that we must learn to ask the right questions, and that is what we intend to do. Generally, we will try to avoid meaningless questions. Many will object to this, saying that the controversy does not go away by refusing to acknowledge questions for which we have no answer. And, besides, who shall decide what is 'meaningless'? Nevertheless, if quantum mechanics is silent on an issue, then so will we. We will try to stay focused on results and probabilities.

Along with Bohr, we trust that quantum mechanics is a complete theory. It tells us all there is to know about an atomic event. If we stick to the experimental facts, then we know all there is to know and there is no need to speculate about anything else.

If we do take quantum mechanics at face value, we are forced to concern ourselves only with the results of experiments. Most agree that classical physics fails to explain how a quantum experiment works. There is no quantum explanation either. In particular, it is generally believed that there is no physical process that brings about the experimental result. There are no inner workings to concern us. But, many physicists, Einstein in particular, refused to accept this. They are realists, and, for them, there is more to an experiment than just the results. They know that something real must be going on in the experiment before we obtain the result. Yet, quantum mechanics denies any such underlying machinations.

But, the realist argues that we construct experimental apparatuses from real devices that are part of our macroscopic world. And, part of our job as physicists is to describe how that experiment works. Bohr and his followers, on the other hand, believed there is no explanation of how the quantum experiment works and it is futile to think otherwise. According to Bohr, the different parts of the experiment lose their individuality and become entangled into a single 'wholeness'.

The quantum experiment is not a collection of its parts, in spite of what we see in the lab. Rather, it is a 'whole' thing, a single entity, an 'entangled' quantum thing. There is no classical analog.

This inseparability of its parts is the defining characteristic of the quantum experiment that makes it radically different from its classical counterpart.

Quantum theory suggests the world, at its most elementary level, is probabilistic, not deterministic as Einstein wished it to be. Experimental results are generated without cause in a purely random way. There is no answer to the question, "How did that happen?" As difficult as it is to accept, any questions about an underlying atomic reality are pointless, because there is no 'underlying atomic reality'. In Bohr's world, there are no atoms, only atoms observed under specified experimental conditions. That's all there is!

Thus, atomic particles are not real in the classical sense. They do not have a separate existence of their own. Quantum particles do not have individual identities. In particular, they are not separate from other parts of the experimental apparatus used to make measurements on them. Rather, they are part of the experimental whole, not independent of it.

And, most importantly, a quantum experiment must always have a result. We must finish the experiment. There is nothing to discuss if the measurement result is lacking. As we have said, quantum mechanics is all about results. 'No result', then, means 'no experiment is done'. If we want to 'see' a quantum particle, we must do the experiment and obtain a result. We must 'look at it'!

> *The quantum particle 'is there' only when we have an experimental result that tells us that the particle 'is there'.*

Einstein never did accept this. For him, a particle is always 'there', even when we don't know where 'there' is. But for Bohr, the only particle is the observed particle.

Einstein once asked a colleague if he, "really believed that the moon only exists when you look at it"? This is not a hard question for the classical physicist to answer. The classical physicist is a realist. He knows that the moon is real and it doesn't matter whether or not any one is looking at it. Of course, it is always there! If you have any doubts about this, then, by all means, go outside and look. Ever since man first looked up to the heavens, he has seen the moon. It is safe to say that billions of people have seen the moon throughout the ages. From different locations, at different times, the moon has always been there, just where the classical laws of motion say it should be.

We are so sure of this that we construct calendars based on the moon's periodic motion. And, we can use the known position of the moon for navigation and to predict the tides.

It doesn't make any sense to ask such a question. It is a well established fact, as well as anything can be, that the moon is real. And, like all real things, it is always somewhere at every instant, even when no one is looking at it. Why, then, would anyone doubt the moon's existence when no one is looking?

Einstein asked the question because quantum objects do not behave the same way the moon does, as much as he would have liked it to be otherwise. Atoms are not little moons. Bohr argued that quantum theory involves particles that are quite different from the moon and we have no right to expect the laws of Newton and Maxwell to extend all the way down to the atomic level.

Simply put, classical physics does not work in the quantum domain. We must accept quantum mechanics, then, as a new, albeit extremely counterintuitive, description of nature that is necessary if we are to investigate the atomic world. But, if we refuse and instead insist that quantum objects are simply smaller sized classical objects moving in space-time, along with everything else, then there is no hope. We will be forever confronted with all kinds of weird behavior. One only needs to consult the literature on this subject to become completely bewildered.

For example, much has been written about Schrödinger's hapless cat, the most famous cat in all of physicsdom. As mentioned above, some say that this cat has been seen alive and dead at the same time. Others argue that it is half dead and half alive, whatever that means. Weird behavior, indeed!

1.3 Schrödinger's Cat Experiment

We will use the Schrödinger cat experiment as an introduction to the approach taken through most of the book where we give a classical analysis and then compare it with the quantum approach. It will also serve as our introduction to experiments that have only two possible outcomes.

Experiments that measure these so-called 'two-state variables' are among the simplest to analyze, yet, they possess all the counterintuitive quantum attributes of more complicated systems. We will employ such cat-like experiments throughout the rest of the book.

In 1935 Irwin Schrödinger published a paper entitled, "The Present Situation in Quantum Mechanics". It has become famous for just one paragraph, in which he described a ghastly experiment involving a cat whose name shall forever be linked with quantum weirdness.

Schrödinger reported that he had confined his cat to a closed box containing a radioactive nucleus connected to a lethal device. There is a fifty-fifty chance that the nucleus will decay within one hour and emit radiation that triggers the killing apparatus. (Relax! This is just another made up thought experiment. As far as I know, Schrödinger always took good care of his pets, including his famous cat, who, I am told, lived to a very old age).

When closed, the box is sealed in such a way that there is absolutely no way to know if the lethal device is on or off. We have no access to the interior. It is truly impossible for us to determine anything about what is going on inside the closed box. We know nothing of the cat until we open the box. Before the box is opened we do not know if the cat is alive or dead. This is the nature of quantum experiments and, obviously, not the way of ordinary boxes.

When we do open the box, we always find a decayed nucleus with a dead cat or we find a not-decayed nucleus with a live cat. These are the only possible outcomes of this experiment.

We get one of those two results. Only one of the two! We never find the cat to be alive and dead at the same time. And, rest assured, no one ever will. Likewise, the nucleus is never decayed and not-decayed in the same measurement.

If we repeat the experiment many times, in half of the experiments, the cat survives. The other half of the experiments yields a dead cat. These are the facts concerning this experiment.

Schrödinger was an Einstein sympathizer, even though he had spent time with Bohr in Copenhagen. Apparently he could accept the probabilistic description of the decaying nucleus, but extending its indeterminate nature to the macroscopic cat was, in his words, "quite ridiculous". He wanted his cat experiment to be an exercise in absurdity. To this day, many believe that he succeeded.

1.4 The Classical Description of the Schrödinger Cat Experiment

Einstein, the classical physicist, might ask, "Is the cat dead or alive after one hour?" He doesn't care if the box is open or closed. And, he expects a clear-cut answer. His kind of physics provides definite answers to such questions. He wants to be told, "The cat lives". This can then be confirmed by repeating the experiment and always finding a live cat. Or, if it should happen, "The cat is dead." Repeat the same experiment and again the cat is dead.

Every time we run the experiment, there is a 50-50 chance that the cat is found alive. This probabilistic statement might very well be true, but the classical physicist believes that it is not the entire story. At any time during the hour, the cat is still a real thing and, as such, it is definitely alive, or it is not. It is alive until the nucleus decays, if that should be the case. Then the lethal device is turned on and the cat dies.

Einstein wants to discuss the plight of an ordinary cat in an ordinary box. After all, Schrödinger himself said that he put it there. For the classical physicist, all cats are real and Schrödinger's cat is no exception. This is a real cat in a real scary situation.

Real cats obey the deterministic laws of classical physics, so there must be a classical explanation of how the cat survives, or why it doesn't.

According to classical theory, identical experiments always give the same result; a classical experiment has a definite result. There is only one possible outcome. Doing the same experiment over and over always gives the same result. It is determinate.

But, the Schrödinger cat experiment has two possible results, not one. It does not matter how carefully we construct the apparatus and perform the experiment. If we repeat this experiment many times, sometimes the cat lives and sometimes the cat dies. We have no control over the outcome. Even if we could do a perfect, error free experiment every time, we still would not always get the same result. Here, it seems, the same experiment has two possible outcomes.

"Not so!" says the realist. "There is no single classical experiment that will sometimes kill the cat and at other times do it no harm. The experiment that kills the cat is obviously different from the one that does not. There must be two different experimental procedures at work here, one for each result."

Classically, then, the Schrödinger cat experiment actually defines two different, but each deterministic, experiments. There is a 'live' experiment in which the cat always lives and there is a 'dead' experiment in which the cat always dies. But we don't know which experiment we are doing until we open the box.

Whether the cat lives or dies depends on which experiment is done. As in all classical experiments, it is the inner workings of the experiment that determine the outcome. Classically, we are able to describe the process that leads to each result.

In the 'live' experiment, the unstable nucleus never decays and the lethal device is never turned on. The cat lives. If we repeat this same experiment many times, we always get the same result. The cat is alive when we put him in the box and he is alive one hour later when we open it. Every time! Classical experiments are like that. The cat must have been alive and unharmed the entire hour inside the closed box. Cats have been known to nap or frolic in boxes without any ill effects. So it must have been in the 'live' experiment.

But, in the 'dead' experiment, a nucleus does decay and the killing device is triggered by the emitted radiation. When we open the box, we find a decayed nucleus along with the dead cat. Every time we do this particular experiment, we find a dead cat. Common sense tells us that the cat was already dead when the box was opened. It would have died shortly after the nuclear decay that turned on the lethal device.

Each experiment, being classical in nature, is deterministic. We can predict the outcome. No pussyfooting around with probabilities here! Einstein wants a definitive answer, and each experiment gives it to him.

The problem with this classical approach, however, is that we never know which experiment we are dealing with until we open the box. But the fate of the cat is actually determined by whether or not an unstable nucleus will decay in one hour. Since live cats and undecayed nuclei are always found together, we must ensure that only those nuclei guaranteed not to decay take part in our 'live' experiment.

Unfortunately for the cat, there is no way for us to do this. We have no control over the decay of unstable nuclei. It is impossible to predict which nuclei will decay and which will not. But, if we knew more about nuclear decay, then we could explain why one nucleus decays while its identical neighbor does not. We could then use only those nuclei that will not decay. It is our ignorance of the decay process that prevents us from actually constructing the 'live' experiment.

Nevertheless, each time the cat is placed in the box we are doing a deterministic classical experiment with a definite outcome. It just so happens that we don't know what the outcome will be.

Common sense tells us that the cat is in the closed box for the entire time prior to opening. It doesn't matter that we cannot see inside the closed box. The classical physicist believes in live cats and in dead cats. He believes that being alive or being dead has nothing to do with whether the box is open or closed.

Classical physics is able to predict the result of each measurement along with the corresponding procedure that brings about each outcome. We know why each result occurs. But we cannot predict the result of an individual experiment because we don't know which experiment we are doing. If we did know, then we could predict the result with certainty. This ignorance of which experiment is actually being done forces the classical physicist to accept a probabilistic description. But, he is confident that once he obtains a complete understanding of the nuclear decay process, there will be no need for probabilities. It is our lack of controllability over nuclear decay that gives rise to the probabilistic nature of the Schrödinger cat experiment.

For the classical physicist, there is nothing strange about these experiments. Unusual things do not happen to cats in boxes. But it just so happens that we are ignorant of some important details. Nevertheless, there is always a reasonable explanation of what is going on.

But, the claims of quantum physicists are another matter!

1.5 The Quantum Description of the Schrödinger Cat Experiment

We now assume that Schrödinger's cat obeys the rules of quantum mechanics. Bohr would argue that the Schrödinger cat experiment is actually one experiment, a single, obviously non-classical, experiment with two possible results. He accepts it as counterintuitive and indeterminate. There is no definite answer to the question, "Is the cat dead or alive after one hour?". We only know that when we open the box, in 50% of the experiments done the cat lives, and 50% of the time the cat dies.

The theory does not describe anything beyond this. If the theory is complete, then there is nothing else to describe.

In the quantum description we do not concern ourselves with what is happening to the cat itself. Rather, we describe an indeterminate experiment done with a cat, quite different from an ordinary cat. This indeterminate description is necessarily statistical. We do not know what result we will get until we open the box, but we can predict its probability of occurrence.

There is no 'live' experiment and there is no 'dead' experiment as the realist claims. Rather, the 'live' experiment is identical in every way to the 'dead' experiment. They are one and the same. But in the quantum experiment there is no way, even in principle, to determine which result we will get when we open the box. We only know that half of the time we find a live cat and half of the time we find a dead cat.

Quantum mechanics is not a science of definite outcomes. The two possible results and the probability for each possible result are the only kinds of information obtained from the quantum cat experiment. This is typical of quantum experiments, where there is more than one possible result. Its probabilistic nature is inherent in the experiment itself, and it is not a consequence of our ignorance of the nuclear decay process.

Bohr says that this quantum description is complete. It tells us everything there is to know about Schrödinger's cat experiment. We know the two possible results and the probability of obtaining each result. There is nothing else!

There are no inner workings, no hidden reality, and no physical mechanism that is missed by the theory. There is no need to speculate beyond the results and the probabilities. If quantum theory, along with the experimental observations, does not provide any information about what is 'happening inside the closed box', then, there is 'nothing happening inside the closed box'.

There is no detailed explanation of how the cat experiment works and there is no apparent reason why one cat lives and another cat dies. There is no quantum description of what is happening to the cat inside the closed box. Whether the cat lives or dies is a purely random event. It is simply the luck of the draw!

But, quantum mechanics does provide a quantum explanation, of sorts. Schrödinger's cat experiment is indeterminate because Schrödinger's cat is no longer an ordinary cat in an ordinary box. The finite size of Planck's constant introduces a 'wholeness' into the quantum experiment not seen in classical physics. The different parts of the cat experiment are no longer separable. Rather, all parts constitute an inseparable whole.

Now, it is a quantum cat, a cat-in-a-box, and the experiment is governed by the statistical rules of quantum mechanics. A quantum cat-in-a-box is an entangled thing where the cat and all the experimental parts make up a single entity. There is nothing like this in our everyday world. We can't imagine such a thing! Whereas the different parts of a classical experiment are distinct real things, all running in the expected way, the different parts of a quantum experiment are not separable.

The cat, the box, and everything in it, including the unstable nucleus and the killing device have lost their individual identities. They no longer exist on their own. 'Entangled' means that all the parts now make up an inseparable wholeness where all the parts have a kind of shared existence. We should not be talking about a cat at all. But, we do, and that is the source of much of our confusion.

In the quantum description there is only the cat-in-the-box, the entangled thing. When we say 'cat-in-a-box' we are referring to the entire entangled apparatus, which is what quantum mechanics is about. Quantum mechanics does not apply to the solitary cat, only to the entangled whole cat-in-the-box.

If this is truly a quantum experiment, then there is no sequence of events that triggers the lethal device, as suggested in the classical description. That sequence is a classical process. It is an illusion! There is no such thing as a quantum 'sequence of events'. Even though we designed and built the apparatus to function in an ordinary way, it does not do so.

Quantum mechanics is silent about any happenings in the closed box. There is no 'live cat in the box' until we open the box and actually see a 'live cat in the box'. Likewise, there is no dead cat anywhere until we open the box and see a dead cat.

Some misinterpret this to mean that we kill the cat just by looking at it. This implies that the cat was alive just before opening the box. But, we have no experimental evidence of that and quantum theory makes no such claim. Further, there is no process that allows the cat to live, or one that kills the cat, if that be the case.

The point is this - quantum mechanics does not describe the behavior of a cat inside the closed box or the process that kills it. Besides, we have all observed cats in various situations and, as far as I know, no one has ever killed a cat by merely looking at it. That would require supernatural powers.

But, why a cat? While we are at it, let's make it an elephant. Since this is only a thought experiment, a mental exercise that does not involve hurting any real cats, or elephants, for that matter, it is just as easy to think of an elephant as a cat. And, killing an elephant simply by looking at it would be much more spectacular!

"This is all nonsense", cries the realist. "After all, everyone knows that the cat is always in the box, even when it is closed. It was a cat when we put it into the box and it was still a cat when we opened the box one hour later. The cat had to have been in the box the whole time. Why should we think otherwise? That's common sense, isn't it? An acceptable theory would describe a real cat in a real box, even when the box is closed. It would provide more than just what to expect when the box is opened."

Einstein does provide a reasonable explanation of why the cat lives, or why the cat dies. He knows the inner details of the experiment. Bohr, on the other hand, says there is nothing to explain.

Of course, this is just a thought experiment that has never been done. We have discussed it because of its notoriety.

We have used the cat experiment as an introduction to the differing points of view of Einstein and Bohr. It contrasts the separability of classical theory with the non-separability of quantum mechanics. We will see this over and over again as we transition through this book.

For Bohr, there is no real cat for physicists to fret over. There are only the results of an experiment done with a cat. The quantum description correctly tells us what to expect when the box is opened. But all the underlying details found in classical description are missing.

Many of us want to know what did happen to the cat prior to opening the box. We are not yet ready to give up on classical physics. After all, it makes the most sense and it has served us well in the past.

And, besides, all this entanglement talk is extremely counterintuitive and there doesn't seem to be any need for such an extreme view.

The real source of our confusion is the fact that a real cat is not a quantum thing and we cannot make it so. In our real world, it is impossible for a cat to become entangled with a box and all the other apparatus. The cat is not atom-like. The extremely small value of Planck's constant can be ignored at this level. Classical physics is perfectly adequate to describe a cat, a solitary cat, in a box.

Einstein is the common sense guy in all of this, and consequentially, his presentation is the easiest to understand. He speaks our language! But, physics is an empirical science and it is not how reasonable the theory might sound that will win the day. Ultimately, we must do authentic quantum experiments similar to the cat experiment that will allow us to determine which of the two descriptions is correct. Only then can we say that atomic objects are separate and real or that they are entangled and not real. We will do this soon enough.

But, there are other things that we must do first.

2

THE CLASSICAL UNIVERSE
The world view of Albert Einstein

The physical universe is real.
          Albert Einstein

2.1   General

Ultimately, we will discuss quantum mechanics. After all, that is the purpose of this book. But, before we do that, we must introduce you to its predecessor, classical physics. This is necessary because much of the confusion surrounding quantum theory has its source in the classical theory, which is the physics of things familiar to us. We are used to having things happen in a certain way, the classical way.

We are comfortable with classical physics because it describes common place behavior. All the things around us obey the laws of Newton, Maxwell, and Einstein. We live in the classical world.

We all look at the universe through classical eyes. We are a part of it. Thus, our study of quantum mechanics is biased from the onset by our classical experience. Quantum mechanics appears strange to us because it is not what we have come to expect.

So, we begin our journey with a purely classical event. Although not an everyday occurrence, it has all the features inherent in our classical description of nature.

2.2   The total solar eclipse of 21 August 2017

At near dawn over the Pacific Ocean, at 16:48:33 UT (Universal Time), about halfway between the Hawaiin islands and Alaska, on Monday, August 21, 2017 the moon and the sun rose together, the moon totally eclipsing the sun. It was a spectacular sight, witnessed by only a lucky few. As the moon-sun rose, the moon's shadow sped across the Pacific towards the United States, making landfall near Newport, Oregon at 17:15:50 UT, about 10:16 local time. The total eclipse was visible in this area for about one minute and fifty seconds.

For the next one hour, 33 minutes and 17 seconds, the moon's shadow raced across the United States in a southeasterly direction at a supersonic speed of about 2000 mph with a path width of 60 miles or so. Hopkinson, KY had the distinction of having the longest lasting total eclipse. They were in the dark for 2 minutes 40 seconds.

The last place in the United States to view the total eclipse was near McClellanville, South Carolina where the shadow passed into the Atlantic Ocean at 14:49:07 local time. The total eclipse eventually ended over the open ocean at 21:04:30 UT, 4 hours and 6 minutes after it began in the Pacific. It never reached Africa.

Amazingly, all this information was known to us years before the event actually occurred!

What kind of magic do we possess that gives us the power to determine such things long before they actually occur? It is not magic, of course. Rather, we can thank Sir Isaac Newton for his laws of motion and for his theory of gravity. The mathematical formulation of Newtonian mechanics is in terms of the calculus, also invented by Newton. (Actually, most agree that Gottfried Leibniz also developed calculus at the same time, but independent of Newton, although there is still some controversy over who actually invented what first.) The future motions of the earth, moon, and sun can be predicted by these mathematical laws.

To make such predictions, we must apply Newton's laws to the individual motions. Actually, we want to determine the path taken by the moon's shadow across the surface of the earth as the moon passes between sun and earth. To make our calculation more difficult, the shadow falls on a spherical earth rotating about an axis that is tilted to the plane of the earth's orbit around the sun. Further, the moon's orbit around the earth is not quite in the earth-sun plane. Not an easy problem. It is a brutal calculation if you were to try this longhand. Thankfully, no one does!

The predictive powers of such mathematical laws are certainly evident in our description of the 2017 eclipse. The mathematics told us exactly where the moon's shadow would be at every instant, and we knew how big it would be. There were many excellent web sites that told us where and when we could witness this extraordinary event. And everything occurred exactly as predicted. We knew that all of North America would experience at least a partial eclipse. People knew what to expect at their locations. All this from classical physics.

But, wait! There is more. (Is this beginning to sound like a TV commercial?) If we look more deeply into the equations, we discover that the 2017 solar eclipse is just one event in a periodic cycle. We find that an eclipse has occurred every 18 years and 11 days beginning with a partial solar eclipse on January 4, 1639.

A total eclipse was not seen until June 29, 1927 and the final total eclipse will occur on September 9, 2648. The entire series will end with a partial eclipse on April 17, 3009 after a total of 77 such events. The 2017 eclipse was the 22nd in the series.

Every eclipse in the series has been cataloged. We know the exact time that each eclipse begins and we know the exact path taken by the moon's shadow across the earth's surface. We know when and where each eclipse begins and we know where it ends.

Everything that we could possibly want to know about any eclipse in the series is contained in the mathematical equations. The earth, moon, and sun are real objects that are part of the classical universe. They interact with each other according to the gravitational laws of Newton, from which we determine the trajectory of each object. The equations predict when an eclipse will occur, and why it occurs is explained in terms of the classical laws.

We have presented all this information, some rather minute, because we wanted to emphasize the exhaustive kind of detail that can be extracted from the classical laws. The equations discovered by Newton provided us with everything there is to know about the solar eclipse of 21 August 2017. Ask a question of it and Newton's equations will provide the answer.

The classical laws not only provided predictions about the eclipse, but they also provided an explanation of why and how it all came about. According to Einstein, this is what physics is supposed to do.

Classical physics has to be one of the greatest triumphs of the human mind! It is everything that a science should be. It provides us with a complete description of the physical world. There is no need to speculate about what is going on. The classical laws of Newton and Maxwell provide all the relevant details.

It is understandable why so many are reluctant to give it up.

2.3 Classical physics is about reality

At some point in our existence, human beings began to ask questions about their environment. We interact with our surroundings using our senses. We learn to identify different objects using our sight, hearing, smell, touch, and taste. We see objects moving about and begin to ask questions such as, "Where is the tiger?" Sometimes, it is crucial that we know the answer to such questions. If we are in a forest with a tiger, it is important to know where the tiger is. "Is it moving? How fast is it going? Is it coming towards us?" These are important questions. Early man needed to know such things in order to survive.

Questions were then asked of other things that were not as threatening. We are inquisitive beings. We wanted to know what was going on around us. This eventually led to doing experiments and then to developing theories to explain the observed results. This development evolved into what we call classical physics.

Classical physics describes the real world of our everyday experience, the universe that we can sense. It is our perception of physical reality.

We have classically conditioned minds that evolved from observing and coping with the world in which we find ourselves. We think in a classical way. We naturally analyze situations based on our experience, which is everything classical, so that we naturally expect all things to behave in the same ordinary way.

We are guided by our intuition and common sense, which have been tempered by the behavior of ordinary things. What we actually observe every day is the classical behavior of things that obey the equations of Newton and Maxwell. What we call 'common sense' is logic based on our classical experience. 'Common sense' is really 'classical sense'.

Our classically tuned brains provide us with mental images of real things we have seen. When we speak of tigers, we have a mental image of the tiger, even when there is no tiger nearby. That mental image is a replica of a real tiger based on actual observation. We use such images to represent real tigers.

Classical physics is a pictorial science. It provides pictures of the real things that make up the physical world.

When we think of an atom, our brain still provides us with an image of a solitary atom, even though we have never seen one. Well, maybe we have seen a picture of one in a book. Maybe not. It doesn't matter. The brain will, nevertheless, generate a picture for us. There is no avoiding this. Unfortunately, in this case, our intellect fails us. The image you have of an atom is wrong! Atoms are not part of our real world experience and our brains are not capable of imagining what an atom 'looks like'.

2.4 Objective reality

Einstein believed that "there is a real world out there". Most of us would agree. After all, we experience it every day. This is the objective reality studied by classical physicists. All things in the classical universe are part of that physical reality and the concepts and laws of classical physics refer to those real things. In fact, classical physics requires a real external world for examination. Without it there is no classical physics.

Real things exist on their own with their own attributes and their own behavior, even when we cannot see them. Although we design experiments to measure the different physical properties used in the classical description, the values so obtained do not require that we measure them. A real object has exact values for all its attributes at all times, even when those values are unmeasured and unknown to us.

Physical properties, such as position and momentum have definite values at every instant, independent of any measuring process. When we do make a measurement, we obtain the true property value possessed by the object just prior to the measurement. In classical physics, measurements reveal pre-existing property values that are present whether or not we do the measurement.

Many times, experiments that measure different physical properties cannot be done simultaneously. They are mutually exclusive. Sometimes it takes many different experiments, done at different times, to learn all there is to know about an object. Taken together, the different observations provide a more complete picture of the object under investigation.

We emphasize that the classical universe does not depend on us looking at it. Objective reality means that the observed universe exists as it is without us. Its existence does not depend on physicists and their instruments of observation. The physical universe is independent of any measuring devices used to observe it.

For example, when a real tree falls in a forest, it does make a sound, even if there is no one around to hear it. A real falling tree is a thing that exists independently of other things, including us. The falling tree and the sound it generates are real things that have nothing to do with what physicists might be doing at the time. The existence of sound does not need someone to hear it or any instrument to record it. It is there by itself, free of the other things in the forest.

Classical physics describes an undisturbed universe that does not require any observing experiments for its existence. A real object is always out there, even when no one is looking at it. Our description of a real object, or a real attribute, is the way we imagine it to be without any interference from us. The classical universe is not bothered by physicists poking at it with their instruments, and everything works just fine without us.

2.5 Real things exist in space-time

The classical universe exists in the continuum of space-time. Obviously, a real object exists somewhere at every instant. Thus, a moving real object has a trajectory. We can take pictures, even moving pictures, of real objects as they are.

The classical laws provide the necessary space-time description from which we can predict the trajectory for each real object in the physical universe. If we know the position and the momentum of an object at any moment, then the equations determines its trajectory, which reveals where it has been, and where it is going. This was most evident in our discussion of the 2017 eclipse, where Newton's laws predicted the trajectories of the sun, earth, and moon separately. Each had its own trajectory given in terms of its instantaneous position and momentum. In particular, we knew where the moon's shadow would be at every instant from beginning to the end of the eclipse.

This is true for all real objects. But if we do not know the initial values of position and momentum, then we cannot apply the classical equations. For example, the moon and a tennis ball are real objects moving through space-time. Each has its own trajectory correctly predicted by the classical equations. The moon and tennis ball each move in a predictable path under the influence of the Earth's gravity. The moon's trajectory is well known. We made use of it in describing the 2017 eclipse.

But, the path taken by a tennis ball is not as certain as we would like because we usually do not have good control over its collision with the tennis racket. Nevertheless, like all moving real objects, it does have its own unique trajectory determined by Newton's laws of motion. Unfortunately, we don't have enough information to make use of them.

2.6 The Locality Principle

Locality and separability are consequences of real objects existing in space-time. All real objects are subject to the principles of locality and separability. These are the distinguishing characteristics of space-time and, therefore, of classical physics. Locality and separability taken together is called local realism.

The locality principle states that in order for a moving object to get from point A to point B, it must pass through every point on the path taken. A moving object passes through the continuum of points that make up its trajectory. Therefore, trajectories are necessarily smooth. There are no jumps or jerks or discontinuities as it moves from point A to point B. It is impossible for an object at point A to suddenly appear at point B without ever passing through all the points in between. Such non-local behavior does not occur in the real world. The classical universe runs smoothly because real objects exist in a space-time continuum, where locality is the rule.

A consequence of locality is that all forces acting on an object are necessarily 'contact' forces that are present at the location of the affected object. Exerting a force on an object at point A can have no immediate effect on an object at a far away point B. Sticking a pin into a voodoo doll has no harmful effects on anyone some distance away from it. That would require 'action-at-a-distance', a non-local interaction that is specifically forbidden by the locality principle. Voodoo dolls do not work in Einstein's universe. The real world of classical physics is free of such behavior.

Furthermore, all local interactions are restricted by Einstein's special theory of relativity, which requires that any signal sent from point A to point B must travel at speeds less than the speed of light. When we say 'signal' we mean the transport of mass and energy.

It is possible for things to move at speeds greater than light speed, but those things cannot carry mass or energy. Shadows, for example can move at superluminal speeds, but electrons cannot.

2.7   The principle of separability

Classical realism is also characterized by the principle of separability, which says that a real object has its own location in space-time. Different objects are necessarily separate from one another. An object exists independently of all the other real things spatially separated from it. Different real things each has its own identity, its own attributes, along with its own location in space. If we remove one of the objects, the others will endure without it.

The mutually independent existence of spatially separated things makes it possible for the physical laws to be applied to each individual thing, apart from other things. When discussing the solar eclipse of 2017 we treated the sun, moon, earth, and moon's shadow as separate independent entities. We were able to determine the trajectory of each one separately from the others.

Separability allows us to connect all the different parts to build a machine. We know the role of each part and how it contributes to the overall operation of the device. We know how the machine works because we know the function of each individual member. The principle of separability enables us to build things, maintain them, and to repair them, if necessary. In the real world we can figure out how things work because we can examine each part individually.

The separability of real things is evident in the objective reality seen all around us. We cannot ignore it. It is a given characteristic of all real objects. The important point is this – the physical world is separable into all its parts and each part has its own existence in space-time.

Separability and locality make possible the deterministic description that is found in the mathematical laws of Newton, Maxwell and Einstein.

Most importantly, each object has its own physical properties whether or not we are observing it. The values of its physical attributes are separate and independent of any measuring device used to observe them. It follows then, that we can ignore the interaction between the object under investigation and any measuring instruments. In particular, physical properties that are observed in experiments have the observed values whether we measure them or not. Attributes 'belong' to the object with a value not determined by the measuring instruments. Rather, the measuring device records a pre-existing value that was there all along.

For example, our description of the moon is of the moon itself, as if there were no instruments focused on it. The devices used to measure the moon's momentum, for example, are not included in our description of the moon. The observing instruments can be turned off and, nevertheless, the moon will still be there with its own momentum, and with its other physical properties, as well.

The true significance of the separability principle in classical theory became apparent, most emphatically, in 1935, when Einstein, Podolsky and Rosen, argued that if the principle of separability is true, then quantum mechanics is necessarily incomplete. Einstein believed that the science of physics would be impossible without it. For Einstein, objective reality consists of real things spatially separated from each other and obeying the principle of locality.

For Einstein, "Reality is thus the thesis of spatial separability." The many separable parts are always there with their own attributes, even when no one is looking.

As trivial as separability is for us, there is no evidence that atomic things are also separable. On the atomic level, separability is replaced with a non-separability, what we call entanglement, which does not allow for a detailed description of individual quantum things.

In fact, entanglement disables both separability and locality. It is the undoing of the classical concept of objective reality. As a consequence, we do not know how things work on the atomic level.

But Einstein never gave up his belief in objective reality, where locality and separability reign. He believed that if quantum mechanics rejects objective reality, then there is something wrong with quantum mechanics.

2.8 Causal determinism

Everything that happens in the classical world happens for a reason given in the classical laws of Newton, Maxwell, and Einstein. Cause and effect are inherent in the classical description. We know how events occur. Classical physics is a determinate science because real things obey strict mathematical laws. In principle, we can always predict the outcome of an experiment by first doing a mathematical calculation. And repeating the same experiment over and over again, we always get the predicted outcome; identical experiments utilize identical processes that always generate the same result, the one predicted by the classical equations. Likewise, different experiments generally give different results.

In addition to making accurate predictions about the experimental results, the classical laws provide an understanding of the real processes and mechanisms that give rise to those results. Sometimes, that mechanism may be unknown to us, but it is there nonetheless. There is always an underlying reality at work in the classical experiment.

The mathematical laws of physics, then, describe a deterministic real universe where events happen for a reason. They do not occur accidentally.

In principle, all future events are predictable. If, for whatever reason, we do not get the classically predicted value when we make the measurement, be assured, there is a reason. Actual measurements are not perfect. Every measurement device is of limited accuracy. And people are not perfect either. But, in an ideal classical experiment, done with perfect measuring devices, and error-free experimenters, the experimental result is the value predicted by the classical equations.

Further, the predicted value so obtained is the true value of the physical property being measured. There is no uncertainty in the predicted value of a real attribute. Probabilities are not inherent in the classical description.

Yet, there are times when we do not know what is going to happen. Not everything is predictable. But when deviations from the predicted value do occur in classical experiments, it is not the fault of the theory. Such deviations are not consistent with classical physics, which always predicts exact values for experimental results. Where errors occur, they are always caused by human ignorance and the limitations of the measuring devices.

Even when we are unable to predict the outcome of an experiment and can only predict the probability of occurrence, the classical deterministic laws are still at work. All classical events are determinate, but, sometimes, our ignorance prevents us from applying the classical laws. Things can get too complicated to use the equations.

For example, when we toss a coin, we cannot predict the outcome. We only know that there is a 50% chance of getting heads, and a 50% chance of getting tails. That is the best that we can do in this case. But, this uncertainty in the outcome is not the fault of Newton's laws.

The deterministic classical laws are always in effect, but sometimes we don't have enough information to make use of them.

Since a coin is a real thing, a coin toss is deterministic. Once the coin leaves the hand, the outcome is predetermined by gravity and Newton's laws. There is a definite outcome to every coin toss.

Unfortunately, we cannot control how the coin leaves our hand, and uncontrollability goes together with unpredictability. We are forced to settle for a probabilistic description.

Nevertheless, there are certain initial configurations of the coin as it is thrown that always lead to a 'heads' result. Unfortunately, we do not know what those configurations are, and if we did know, we still would not be able to replicate a 'head configuration' at will.

But, if we could accurately control the toss every time, then we would get heads every time, if that's what we wanted. We know that the same toss always gives the same result, but, here, we do not have sufficient control over the initial conditions to make any determination. It is impossible for us to reproduce the same toss over and over again. Nevertheless, Newton's laws still determine the outcome. Every time!

In summary, Einstein's realistic world is inhabited by real objects that exist in the space-time continuum. The principles of separability and locality are the dominant features. Real things obey the deterministic laws of classical physics, which provide the required space-time description. As a consequence,

  a. A moving real object has a trajectory.
  b. A real object is separate and independent
    of measuring devices used to observe it.
  c. There is a reason why events happen.

All of this is missing in the quantum description, where entanglement is the essential feature and probabilities are fundamental.

2.9 The classical single-slit experiment

We want to discuss one more classical event. It is among the simplest applications of Newton's Laws. Although the mathematical calculation is trivial, the classical nature of the event is clearly evident. And, that is what we want to emphasize.

Consider a particle that moves in a force free environment, possibly a bullet fired from a gun. The bullet leaves the muzzle of the gun with a given velocity. Newton's First Law states that when no forces act on the bullet it moves with constant speed in a constant direction; the bullet travels in a straight line with constant momentum.

This experimental apparatus consists of the gun, a wall with a slit in it, and a distant screen containing the target. The bullet leaves the gun and goes through the slit without hitting the sides on its way to the screen. The bullet is not deflected as it passes through the slit. In the real world, deflection requires a deflecting force, and, here, there is none. Thus, the bullet has a straight line trajectory and it hits the target at $\theta = 0°$, as shown in Figure 2.1.

If we repeat this experiment many times, we always get the same result; the bullet always travels in a straight line and strikes the screen at $\theta = 0°$. With no forces acting, the particle is never deflected at any other angle, just as predicted by the deterministic laws of Newtonian mechanics.

There is nothing probabilistic about this experiment. Newton's Laws allow us to predict the exact outcome. We know where the projectile will hit the screen. Every time! Without error!

THE CLASSICAL SINGLE-SLIT EXPERIMENT

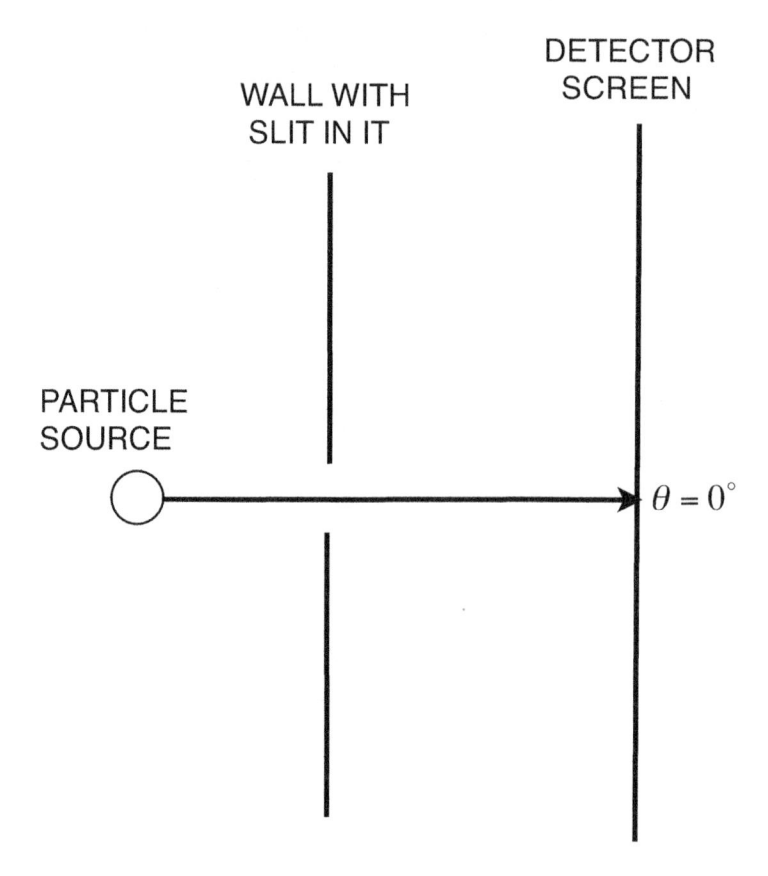

Figure 2.1. A particle, possibly a bullet, is not deflected in passing through the slit. With no forces acting, the bullet has a straight line trajectory, shown as an arrow, from the source to the detector. The particle always hits the screen at $\theta = 0°$ in agreement with Newton's first law. Repeating the experiment always gives the same result.

The motion of the bullet is also causal. Where it goes does not happen by chance. We know why the bullet behaves the way it does. If the bullet is deflected at some angle other than $\theta = 0°$ and misses the target, then there is a reason why that happens. Perhaps the position of the gun was moved. Or, there was an unexpected deflecting force, maybe a gust of wind, that knocked the bullet off course.

The point is this - there is a reason why the bullet hit the screen at $\theta = 0°$ and, likewise, there is a reason why it does not, if that should happen. If necessary, we can make the required corrections so that the bullet again hits the target at $\theta = 0°$.

In classical physics, there is always a reasonable explanation of why something does or does not occur.

Notice that we are describing the motion of the bullet itself, ignoring the rest of the apparatus. The bullet always has a straight line trajectory, with, or without the target in place. In fact, we can remove the wall with the slits and the screen with no effect on the motion of the bullet. The trajectory remains the same. Newton's equations, along with the initial position and velocity, give us the bullets trajectory without any knowledge of the properties of the other parts of the apparatus. For example, the dimensions of the slit and the location of the target have no effect on the motion of the bullet.

3

THE DECLINE OF CLASSICAL PHYSICS
The essential quantum discontinuities make their appearance

I want to emphasize that light comes in this form - particles. It is important to know that light behaves like particles, especially for those of you who have gone to school, where you were probably told something about light behaving like waves. I'm telling you the way it does behave - like particles.
Richard P Feynman

If we are going to stick to this damned quantum jumping, then I regret that I ever had anything to do with quantum theory.
Irwin Schrödinger

3.1 Particles and waves in classical physics

In the real world of our everyday experience, there are two ways of transporting energy from one place to another. We can do it with particles or we can do it with waves. A particle is a localized thing while a wave is spread out in a continuous medium. Classical particles have static properties such as mass as well as dynamic properties such as position and momentum. Particles are discrete objects that can be counted. They have trajectories. A particle moving in space-time carries mass and energy with it.

Waves are not localized in space. They do not have unique positions. Rather, they are spread out over a wide area. Classical wave motion is a way to transfer energy through a vibrating medium without the transfer of any mass. The energy is carried by a disturbance moving through the continuum. That disturbance is what we call a wave. The continuous medium vibrates at a given frequency as the wave passes through. Its frequency is determined by the source of the vibrations, while the wave speed and the wavelength are characteristics of the medium itself.

Classical physics includes mechanical waves and electromagnetic waves. Each requires a vibrating continuum. For mechanical waves, the continuum is a material medium. For example, water waves are mechanical waves where the vibrating medium is water.

Electromagnetic waves are not mechanical waves. The required continuum is not a material medium. Rather, the vibrating continuum is the electromagnetic field itself.

The important thing to remember is that particles are localized in space while waves are spread out. In classical physics there is no controversy about what constitutes a particle and what is a wave. Particles are always particles. Waves are always waves. Particle physics and wave physics are each well understood.

3.2 Light is an electromagnetic wave

In 1803 Thomas Young showed that light passing through a double slit produces bright and dark interference fringes on a distant screen. In classical physics, interference is a property of waves. Since there is no way to explain interference in terms of particles, this was taken to be proof that light is a wave.

At the time, the origin of this wave behavior of light was unknown. That would come later with the definitive work of Maxwell and Hertz. All waves require a vibrating medium, but in 1803 we had no idea what was vibrating as the light wave passed through space.

James Clerk Maxwell later showed that, under certain conditions, the electric and magnetic fields satisfy the same wave equation as mechanical waves. Maxwell's equations indicate that a vibrating electric field always coexists with a vibrating magnetic field to make what we call an electromagnetic wave. Maxwell calculated the speed of his electromagnetic waves to be 3×10^8 meters per second, which was the known speed of light. This led him to speculate that light was such an electromagnetic wave where the supporting medium is the coupled electric and magnetic fields.

After Heinrich Hertz generated radio waves in his laboratory, there was little doubt that light was, indeed, an electromagnetic wave.

3.3 The first mention of 'quanta': Planck's law and blackbody radiation.

By 1900, classical physics correctly described a wide range of physical phenomena. The known universe apparently ran smoothly in accordance with the well established physical laws of Newton and Maxwell. But, there were a few exceptions. In particular, the interaction between light and matter was most perplexing.

By then, it was generally accepted that light was an electromagnetic wave. Maxwell's theory predicted it and Hertz's experiments confirmed it. As a wave, light exchanges energy with matter in a continuous manner. For example, sunlight will burn our skin in a slow, but steady way, if we fail to protect ourselves while at the beach. But, first Planck, and later Einstein, showed that, in some circumstances, light behaved in a dramatically different way.

One of the phenomena that could not be explained by the classical laws was the thermal radiation emitted by a glowing object. All materials, at any temperature, emit thermal radiation in the form of electromagnetic waves. If the material temperature is high enough, the emitted radiation will include light from the visible part of the electromagnetic spectrum.

Supposedly, the simplest case to consider is a perfect emitter, the so called 'blackbody'. Physicists sought to determine how much of each color is present in the emitted blackbody spectrum. Each color is characterized by its frequency or by its wavelength.

It was believed that an explanation in terms of classical concepts should be possible. Why shouldn't it be? Yet, physicists were unable to derive from classical theory the frequency distribution of the thermal radiation that emanates from such a device. The actual frequency spectrum is shown in Figure 3.1.

A small hole on the side of a box absorbs all the visible light shining on it and according to Kirchoff's law, it is also a perfect emitter. Thus, a small hole in the side of an oven is a blackbody.

We heat the oven until the hole glows. We want to predict how much of each color is present in the radiation streaming out of the hole at a given temperature. This should be the same as the radiation inside the heated oven.

The physics problem, then, is to determine the intensity of each frequency (color) inside the oven at a constant temperature. This is essentially a problem for thermodynamics.

BLACK BODY RADIATION SPECTRUM

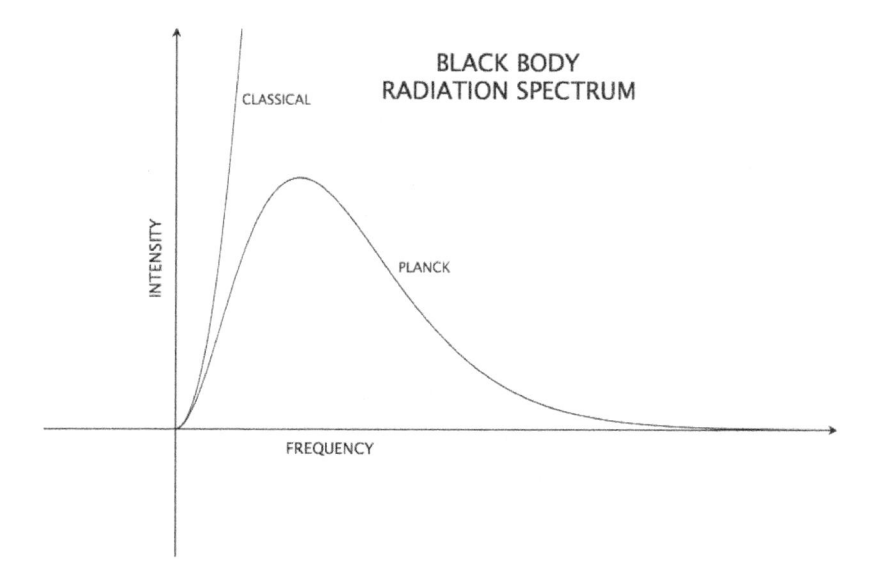

[Graph showing INTENSITY vs FREQUENCY with two curves labeled CLASSICAL and PLANCK]

Figure 3.1. Here we see the amount of each color, given as a frequency, present in the black body radiation spectrum. Planck's radiation law agrees with the observed black body spectrum. Classical thermodynamics predicts that the intensity tends towards infinity as the frequency increases. This is the 'ultraviolet catastrophe'.

The mechanism that causes heated objects to glow was well known before 1900. The theory is straightforward: The walls contain molecules which vibrate when heated. But, molecules consist of charged particles and, according to Maxwell, vibrating charges radiate energy in the form of electromagnetic waves. A vibrating molecule, then, is a tiny radiator of electromagnetic energy. It is a source of electromagnetic waves when it emits energy and it is a sink when it absorbs energy.

In 1900 physicists assumed that a vibrating molecule is like any classical oscillator, like a tiny vibrating spring, perhaps, which can have any value of energy. We can put in, or take out of a classical spring as much energy as we like (as long as we don't break it). When a vibrating spring absorbs energy, it gets stretched to greater lengths. Likewise, when energy is emitted, the spring contracts. In any case, the energy is emitted or absorbed continuously as the length changes.

Using classical concepts, many attempts were made to obtain the intensity distribution of the blackbody radiation, but with limited success. In 1893 Wilhelm Wien discovered his displacement law, which relates the frequency of greatest intensity to the temperature. The higher the temperature, the greater the dominant frequency. The color that we see is determined by the temperature. Thus, we can use Wien's Displacement law to estimate the temperature from the color of a glowing object.

In 1896, Wien also discovered an exponential equation, the so-called Wien's Distribution Law that correctly predicted the spectral distribution for higher frequencies at low temperatures, but failed everywhere else.

In 1900, but before Planck obtained the correct frequency distribution, Lord Rayleigh used a purely classical approach to obtain the experimental results at the lower frequencies. However, his formula predicts an infinite intensity at the highest frequencies. Hence, it came to be known as the 'ultraviolet catastrophe'.

Obviously, there was something very wrong in the classical approach. In the years that followed, numerous investigators, including Planck, attempted to apply different classical schemes to the problem. They all assumed a continuous exchange of energy between the radiation and the inner walls of the oven. And they all ended up with the same ultraviolet catastrophe. Classical calculations could not reproduce the experimental data.

Max Planck succeeded in obtaining the correct intensity distribution by modifying Wien's Distribution Law. But, in so doing, he had to make a most outrageous assumption. He replaced the classical-like radiator with an atomic oscillator that absorbed and emitted radiation discontinuously, in bursts of energy called 'quanta'. Planck assumed that each quantum carried an amount of energy

$$E = nhf,$$

which is now known as Planck's hypothesis. Here, n is an integer and f is the frequency of the vibrating molecule. Planck then adjusted the value of the constant h so that his calculated intensity distribution agreed with the experimental data available at the time. He obtained the value $h = 6.55 \times 10^{-34}$ Joule sec. Sometimes called the universal quantum of action, it is usually called 'Planck's constant' in his honor. The modern value is $h = 6.626 \times 10^{-34}$ Joule sec.

Planck's law predicted the correct intensity at all visible frequencies. When the energy is exchanged discontinuously as quanta, the ultraviolet catastrophe goes away!

Planck's constant has a small, but non-zero value that is responsible for all the non-classical effects that we now attribute to quantum mechanics.

In Planck's hypothesis, his constant relates the quantum of energy with the radiation frequency. But, classically, energy does not have a corresponding frequency! This was entirely foreign to classical physicists of that era. The allowed energies of Planck's atomic oscillator, shown in Figure 3.2, was radically different from an ordinary spring.

Unlike a classical spring, the vibrational energy is no longer a function of its amplitude. In fact, there is no mention of the amplitude at all. Planck's energy radiator has only certain allowed values of energy, which are related to its vibrational frequency, not to its maximum displacement.

This was the first example of energy quantization, where the energy can have only certain discrete values.

In quantum mechanics, a definite energy can be expressed as a frequency.

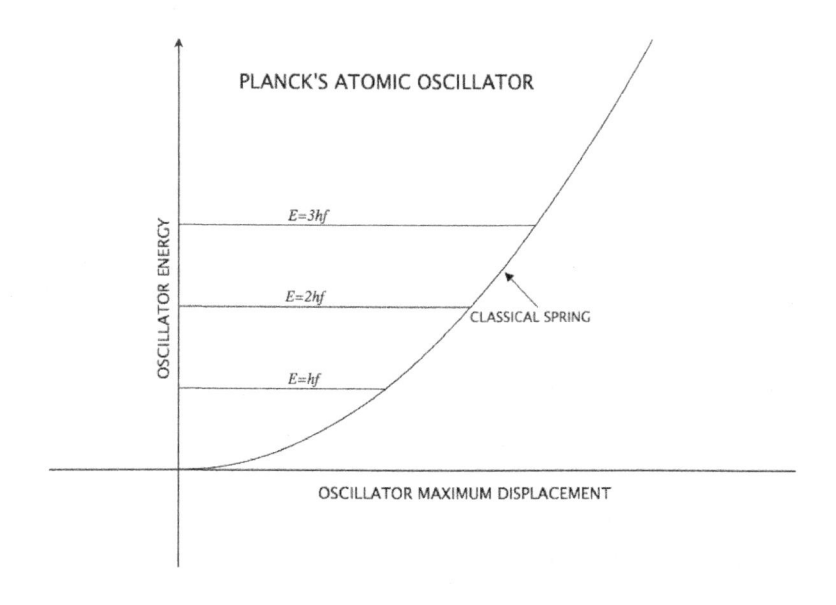

Figure 3.2. Planck's atomic oscillator has only certain allowed energy values, $E = nhf$, shown as the horizontal lines. Here, n is an integer, f is the oscillator frequency, and h is Planck's constant. We show the lowest three energy levels. The smooth curve gives the energy of a classical (real) spring that vibrates with frequency f.

The energy of real springs can have any value depending on its maximum displacement. It has nothing to do with frequency, while the energy of Planck's oscillator has nothing to do with displacement.

As a result, energy is not exchanged in a smooth, well behaved, wave-like manner, as expected. Rather, energy is exchanged between the enclosed radiation and the walls of the oven in discontinuous bursts. The atomic oscillators are not just ordinary vibrating springs as first imagined.

Vibrating springs that possess only certain allowed energies would somehow have to jump from one energy level to another instantaneously when a quantum is absorbed or emitted. Very strange, indeed.

Most physicists of the time refused to accept such a radical departure from classical thought. After all, Maxwell's electromagnetic theory should not be rejected on the basis of this one, albeit successful, ad-hoc equation. Most, including Planck, were confident that a classical explanation would eventually be found. None ever was.

Planck himself made no attempt to attribute any physical significance to his energy quanta. For him, his hypothesis was simply a mathematical means to an end.

For the first time, physicists were confronted with what Bohr would later call an 'essential discontinuity' in quantum behavior. Apparently, the atomic world does not run smoothly. It does not look like the mechanical universe of Newton and Maxwell.

Planck won the Nobel prize in 1918 for "his discovery of energy quanta".

3.4 Light consists of particles: The photoelectric effect

But we still had much to learn about the nature of light. When light shines on certain metals, electrons are emitted from the surface. This phenomenon is called the photoelectric effect and the emitted electrons are called photoelectrons.

The photoelectric effect was accidentally discovered by Heinrich Hertz while doing his monumental work that confirmed the existence of electromagnetic waves.

Classically, we expect that light, being a wave, would, as usual, exchange energy with the metallic electrons in a smooth, continuous way and the energy of the emitted photoelectrons would depend on the intensity of the light. Dim light should give low energy photoelectrons. Bright light would produce more energetic photoelectrons. It all made perfect sense to the classical physicist of 1905.

But, the photoelectric effect resisted any such classical explanation. The experimental fact is that the energy of the photoelectrons has nothing to do with the intensity of the incident light beam. Increasing the intensity increases the number of photoelectrons, but not their energy. And, furthermore, if we increase the frequency, then the electron energy does increase accordingly. More perplexing was the fact that below a certain frequency, no photoelectrons were ever emitted, no matter how intense the light.

Maxwell's electromagnetic wave theory of light had no explanation for these results.

In 1905 Albert Einstein published three papers in the same issue of *Annalen der Physik*. Each was extraordinary. Taken together, they established Einstein as the greatest physicist of the emerging twentieth century. Before 1905 nature obeyed the laws of Newton and Maxwell. After 1905, it was Newton, Maxwell, and Einstein.

One of those papers was his explanation of the photoelectric effect. He embraced Planck's hypothesis that energy is exchanged between light and matter in bursts of energy, but, with a major difference. Planck never said that light was not a wave. He had no intention of challenging all the theoretical and experimental evidence that showed light to be an electromagnetic wave. He only said that energy is exchanged between light and matter in a discontinuous way. But, here, the revolutionary Einstein describes a light beam as a shower of particles, not a wave at all! According to Einstein, Planck's quanta of energy were actually massless particles of light. The light is absorbed one particle at a time. Each light particle is a burst of energy of magnitude hf, just as Planck had hypothesized. Thus, it is Einstein, not Planck, who is usually credited with introducing the corpuscular theory of light. His light particles would later be called photons.

Einstein also assumed that in the photoelectric effect, a single photon gives all its energy to a single electron. The photoelectric effect is truly a 'particle on particle' interaction. There are no waves involved, making it difficult for many to accept the presence of a frequency, which is a wave concept.

In a bright beam of light there are more photons than in a dimmer beam. At a fixed frequency, increasing the intensity, only increases the number of photons and, hence, the number of photoelectrons, with no effect on each electron's energy.

Using conservation of energy, Einstein said that the photoelectron leaves the metallic surface with energy

$$E = hf - \phi,$$

where hf is the photon energy as given by Planck and ϕ, called the work function, is the binding energy that normally holds the electron onto the surface.

Reducing the frequency, reduces the electron energy until $E = 0$ at $\phi = hf_0$, where f_0 is the threshold frequency, below which, photoemission does not occur, no matter how intense the incident light. Below the threshold frequency, there is no photoelectric effect because the incident photon does not have enough energy to kick an electron out of the metal. In such a case, the metallic electron cannot get enough energy to overcome the binding energy. It all makes sense if light consists of particles and if frequency is equivalent to energy. There is no explanation in terms of light waves.

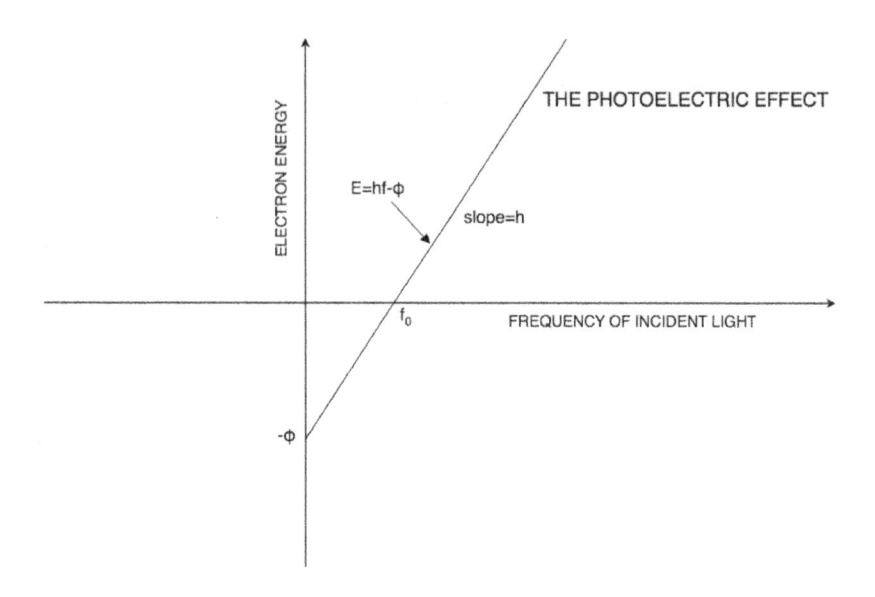

[Figure: Plot showing electron energy vs frequency of incident light. The line has slope=h, y-intercept at -φ, and x-intercept at f₀. Labeled E=hf-φ. Title: THE PHOTOELECTRIC EFFECT]

Figure 3.3. In the photoelectric effect experiment we measure the energy of the emitted electrons as a function of the incident frequency. The incident photon has energy hf and the photoelectron is emitted from the metallic surface with energy $E = hf - \phi$. The experimental plot of the measured electron energy vs incident frequency can be used to get the value of Planck's constant h, the threshold frequency f_0, and the metallic work function ϕ.

In particular, notice that a plot of electron energy vs. frequency, shown in Figure 3.3, is a straight line with slope equal to Planck's constant. Thus, the value of Planck's constant can be obtained from data obtained from the photoelectric effect experiment.

The incoming photon no longer exists after a photoelectric event. It disappears. All of its energy is given to the electron and to the recoiling atom. In fact, a metallic electron that undergoes photoelectric emission must have been tightly bound to its atom. The interaction is actually between the incoming photon and the entire atom, not with a free electron. This is necessary to conserve momentum during the interaction.

Einstein received the Nobel Prize in 1921 "for his discovery of the law of the photoelectric effect". He never did receive the Nobel Prize for his relativity theory.

Later, Arthur H. Compton would consider the scattering of photons from free electrons, much like the way billiard balls are scattered in classical mechanics. Using conservation of energy and momentum, he calculated the loss in energy of a scattered photon. According to Planck and Einstein, there should be a corresponding increase in the wavelength of the scattered light. Compton measured such an increase in wavelength of the scattered light in agreement with Einstein's assumption that light obeys particle physics. This is now known as the Compton effect.

Compton was awarded the Nobel prize in 1927 "for his discovery of the effect named after him."

3.5 So, is light a wave or is it made up of particles?

Modern day experiments leave little doubt that, if we do experiments that allow us to look close enough, light consists of particles. The empirical fact is that we always see photons, never a wave, when very dim light hits a detection screen.

In modern physics labs, we are able to do experiments one photon at a time. Light sources, usually lasers, can be run dim enough so that only one photon, at the most, is ever in the experimental apparatus at any moment. What we see, then, are individual photons making dots on the detection screen.

The detection screen registers a dot every time a photon hits it. We see a dot at the point of impact - one photon, one dot. We say, "we have seen a photon hit the screen", although we have not actually seen a photon. What we see is a dot, localized in space and time, consistent with a single photon hitting the screen.

And, a single dot does not have any wave properties. We cannot discern a wavelength or a frequency from a single dot. If the photon were a wave, as some insist, it would not be localized on the detection screen. We would never see any dots. Very low intensity light waves would still be spread all over the screen. Any interference pattern would fade everywhere as the light is dimmed, but the entire screen will still be illuminated. A single photon hitting the screen is never observed in such a spread out fashion. It always appears as a dot.

Let us do an experiment where we can compare a real wave with a real particle hitting a distant detection screen. Consider a sound wave generated by firing a rifle. The observed sound wave is the 'boom' heard when the gun is fired. We pack the rifle in acoustic insulation, except for the muzzle. So, initially the boom is a localized wave, a wave packet, located at the muzzle of the rifle. Now, the muzzle of the rifle is the source of a localized wave.

But, Huygen's principle states that every point on a wavefront becomes a source of new waves, which spread out in all possible directions. We see this with all kinds of waves. Our boom does just that. It does not stay localized. It spreads out as it moves away from the muzzle.

If our detector screen consists of people standing shoulder-to-shoulder in a circular arc, with the rifle at its center, then everyone hears the boom at the same time. The once localized boom is now smeared over the entire detection area. Everyone hears it. Many different people, at different locations, all hear it. This happens because the boom is a sound wave that naturally spreads out as it moves through the air.

A wave is spread out over the entire screen. A particle is not. Assume, then, that the rifle fires a paint ball, a particle, along with the sound wave. The paint ball strikes the detector at one, and only one, location. Only one person is struck by the ball. Everyone is 'struck' by the wave. There is no doubt that the paint ball is not a wave.

Likewise, the photon is a particle, not a wave. It was defined as such and it is experimentally observed to be always localized. But, we must not think of photons as just tiny, ordinary particles. They are not! They do not behave like tiny paint balls. Although we call them particles, they do not generally obey the classical laws that govern particle behavior.

We conclude that light does consist of massless particles called photons. But, fret not. We shall see that some experiments done with photons do exhibit wave properties.

3.6 Bohr's 1913 description of the hydrogen atom

Other discontinuities began to appear elsewhere in the atomic domain, always accompanied by Planck's constant.

In 1913, Niels Bohr made his first contribution to quantum theory. Ernest Rutherford, of New Zealand, considered by many to be the greatest experimental physicist of all time, had discovered the atomic nucleus while at Manchester by scattering alpha particles from a gold foil. Some of the alpha particles were backscattered in such a way that can be explained only by the alpha particle hitting something much more massive than itself. Rutherford hypothesized that an electrically neutral atom consisted of such a massive positively charged nucleus surrounded by negatively charged electrons of little mass. His atomic electrons orbited the nucleus like a microscopic planetary system.

But, this presented a serious conflict with classical electromagnetic theory. Maxwell's electromagnetic equations do not allow for such a thing. There is no way for an electron to stay in orbit about a positively charged nucleus. In fact, there is no stable configuration, ever, between charged particles alone. A system of like charges is always repulsive and the charges will fly apart. Unlike charges will always be pulled into each other.

Put an electron near a positively charged nucleus and they will accelerate towards each other until they collide. If atomic electrons were ordinary charged particles, they would fall into the positively charged nucleus.

Classically, it is impossible for atoms to consist of negatively charged electrons and a positively charged nucleus, as envisioned by Rutherford. The simplest atom is the hydrogen atom. Its nucleus is a single proton that is positively charged and about 1840 times heavier than the negatively charged electron. If the attractive electric (Coulomb) force is the only effect between electron and proton, there is nothing to stop the electron from crashing into the proton. The atom will collapse.

Nevertheless, in Bohr's original 1913 model of the hydrogen atom he embraced Rutherford's planetary model. Just as the planets are held in orbit by the attractive gravitational force, he postulated that the attractive Coulomb force would be the centripetal force required to keep the electron in orbit.

But, there is a serious difference between planets and charged particles. Any accelerating charged particle continuously emits electromagnetic waves while losing an appreciable amount of energy. An orbiting electron in centripetal acceleration will emit electromagnetic radiation with decreasing orbital energy. It continuously loses energy while it spirals into the proton. It cannot stay in orbit. It is physically impossible. In this classical model, the radiation emitted from the spiraling electron has a continuous frequency spectrum. At every instant, the radiation frequency is the continually increasing orbital frequency of the plummeting electron.

Bohr's innermost orbit in Hydrogen has a radius of $.5 \times 10^{-10}$ meters. A classical electron initially in this orbit would take about 10^{-10} seconds to fall into the nucleus. Atoms consisting of separate electrons and protons do not exist in the classical universe. There is no such thing as a classical atom.

But, why doesn't this happen to the planets? Well, It does! An orbiting planet emits gravitational waves. The difference is that the gravitational waves are about 10^{-34} times weaker than electromagnetic waves and the gravitational energy radiated away by an orbiting planet is negligible compared to its orbital energy.

In spite of such difficulties, Bohr assumed that the hydrogen atom consisted of a single electron orbiting a single proton. (Although the proton was not named until 1920.) But, his orbits were rather special. In spite of Maxwell's equations, they did not radiate away any energy! Bohr did not explain how this was possible. It doesn't make sense, but it worked.

Since Planck's constant has the units of angular momentum, Bohr correctly hypothesized that $\hbar = \dfrac{h}{2\pi}$ was the fundamental unit of angular momentum. An electron's orbital angular momentum, then, had to be a multiple of $\hbar$:

$$\text{orbital angular momentum} = n\hbar,$$

where $n = 1, 2, 3, 4, -----$. No other values of angular momentum are allowed. In Bohr's hydrogen atom orbital angular momentum is quantized. Only those orbits with these special values of angular momentum are possible.

Note that the smallest quantum of orbital angular momentum has the value $\hbar = 1.05 \times 10^{-34}$ Joule-sec, an extremely small number. Such tiny amounts of angular momentum are impossible to observe in our everyday experience. For example, a batted baseball has angular momentum of about 100 Joule-sec, a huge number compared to the Bohr orbital angular momentum.

As a consequence of the quantization of orbital angular momentum, the energy of the Bohr orbits is also quantized.

The orbital energy can only have the values

$$E_n = -\frac{2.18 \times 10^{-18}}{n^2} \text{ Joules},$$

another very small number not seen in the macroscopic world. The energy is negative because it is a binding energy.

But, do atoms ever emit radiation? Of course! If an atom has too much energy it will radiate away the excess - but not continuously as the collapsing classical atom would suggest. Rather, the energy is emitted in discontinuous bursts of energy. Each burst of electromagnetic radiation is a photon of energy $\Delta E = E_n - E_m$. So, the emitted photon energy is the difference between energy levels E_n and E_m, as if the atomic electron is falling discontinuously from one orbit to a lower one.

It was Bohr who first realized that the radiation frequency was not the orbital frequency, as in classical theory. Since there are only certain allowed energy values, there are, likewise only certain discrete frequencies $f = \frac{\Delta E}{h} = \frac{E_n - E_m}{h}$ in the emitted radiation. These are the only frequencies found in the emission spectrum of hydrogen, and of other elements as well. Thus, the characteristic frequencies of the hydrogen atom are discrete frequency values that are determined from its energy level structure. Bohr was the first to do this. His predicted frequency spectrum agreed with values known at the time.

Thanks to Bohr, we are now able to identify a particular atom from its emission spectrum.

Here, we see again that at the quantum level, nature works in a discontinuous way. The atomic angular momentum and energy are quantized, and excited atoms emit photons, not continuous waves. Again, Planck's constant is in the forefront.

But why, then, doesn't the electron keep jumping from one level to a lower one all the way down to the nucleus? After all, we can imagine the atom collapsing in a discontinuous way, from one energy level to the next. It doesn't have to be a continuous transition to oblivion.

Atoms do not collapse because the allowed energies have a minimum value. All atoms have such a non-zero ground state energy, the lowest energy allowed. As a consequence, there are no energy levels lower than its ground state energy. When the atom decays to its ground state, the decay process stops because there is no lower energy level for the atom to fall further into. An atom in its ground state is stable.

Quantum mechanics does allow for stable atoms. Classical physics does not.

We must accept the fact that atomic electrons are not the same as classical electrons. They do not move in circular orbits about the atomic nucleus. Such a model was rejected long ago, but Bohr's obtaining the correct characteristic frequencies of hydrogen brought immediate fame, and a Nobel prize in 1922, one year after Einstein, for "----- the investigation of the structure of atoms and of the radiation emanating from them."

Although Bohr's electron orbits did not survive closer scrutiny, the concepts of 'energy levels' and 'energy states' did. It is now well understood that the emission and absorption of energy is due to transitions between discrete energy levels.

But, we still don't know what the electron is doing in all of this. It doesn't matter. We shall see that quantum mechanics does not provide that kind of detail. Atomic physics works just fine without us knowing.

From the onset, Einstein was troubled by atoms that somehow jumped instantaneously from one energy level to another. It seemed as if an electron in one orbit suddenly appeared in another without passing through the space in between in violation of the locality principle. He had shown in his special theory of relativity paper of 1905 that such non-local behavior is impossible. Others shared his concern.

3.7 Electron capture

But, do atomic electrons ever 'fall into' the nucleus? Sometimes they do. Sort of! There is a nuclear decay process called 'electron capture', which competes with positron decay in certain unstable nuclei. Atomic nuclei consist of neutrons and protons. Some nuclei are unstable because they have too many protons. Such proton rich nuclei become more stable by transforming one of their protons into a neutron. This is usually accomplished by emitting a positron, which is a positively charged particle with the same mass as an electron. A nucleus that emits a positron ends up with one less proton and one more neutron, which is more stable than its original configuration.

It is possible to obtain the same result if the nucleus can absorb one of its own atomic electrons. Some of the innermost atomic electrons of such atoms have a high probability of being in the nucleus where they can be 'captured'. In a nuclear decay, electron capture is equivalent to positron emission.

For example, the unstable, proton rich, nucleus of ^{7}Be (Beryllium seven) is able to absorb an inner atomic electron, changing a nuclear proton into a neutron in the process. The resulting nucleus, ^{7}Li (Lithium seven), is stable.

So, did a Beryllium electron actually fall into its nucleus during electron capture? I have no idea! We should not take this pictorial description of an electron "falling into the nucleus" too seriously. Quantum events do not lend themselves to such a visual depiction. We do not know what happens to an atomic electron that is 'captured'. We only know that what we call 'electron capture' is a real nuclear decay process that does occur in some proton rich nuclei.

4

THE QUANTUM UNIVERSE
The world view of Niels Bohr

An independent reality, in the ordinary physical sense, can neither be ascribed to the phenomena nor to the agencies of observation.

Niels Bohr

Atoms are not things.
    Werner Heisenberg

4.1   Complementarity vs classical realism

Classical physics is an idealization that is valid in the macroscopic world where Planck's constant can be ignored. It is a pictorial representation of nature on the scale of ordinary things, the things we live with, where Planck's quantum, $h = 6.626 \times 10^{-34}$ Joule sec, is an extremely small, insignificant, number.

As a consequence, the discontinuities associated with quantum events are nowhere to be seen. We observe a smooth running universe where all events behave in accordance with the laws of Newton, Maxwell, and Einstein. In principle, nothing unusual happens in our day-to-day experience. Everything has a reasonable explanation.

Quantum mechanics still uses classical concepts to describe atomic events, but in so doing, physical entities and their properties lose their well defined character. Bohr says we are "re-interpreting" the classical concepts when we apply quantum theory to them. For example, the quantum analogs of position and momentum are not the same position and momentum familiar to us from classical physics and they can no longer be defined in an unambiguous way.

In the real world, a moving particle does have a momentum value at every instant, at every location. There is a definite answer to the question, "What is the particle's momentum at position x?". But on the atomic level, if we know where the quantum particle is, then its momentum is completely unknown. The quantum particle does not have a definite momentum at a given position. This is true of some other pairs of variables, as well.

Generally, measuring different physical quantities requires different experimental configurations that are mutually exclusive. In quantum mechanics, Bohr calls the results of such mutually exclusive experiments 'complementary'. Each experiment has nothing to do with the other and we cannot apply the different results in a meaningful way to the same experiment. At least, in quantum mechanics, we can't.

Here, Bohr uses 'complementary' in a way contrary to its usual meaning where different results are used to supplement each other in order to obtain a more comprehensive picture of the system in question. But in quantum mechanics complementary results from mutually exclusive experiments cannot be used in a supplementary way. Rather, they are incompatible with each other.

For example, one of the earliest puzzles that concerned Bohr was the apparent wave-particle duality of light. Young had long ago verified that light is a wave by producing an interference pattern from light passing through two slits. Yet, one hundred years later, Einstein would show that light sometimes behaved as if it was made up of photons, discrete massless particles. How is that possible? Well, it is not possible in classical physics. In the classical world light is an electromagnetic wave that satisfies the classical wave equation. But in the quantum world, whether we observe light as particles or light as waves depends on the experiment done. We no longer describe the behavior of light itself. Rather, we now describe experiments done with light. Light is neither a wave nor a particle, but there are experiments done with light that exhibit wave properties, and there are other experiments that demonstrate particle-like behavior.

The wave experiment and the particle experiment are mutually exclusive. They are complementary experiments done with light.

The wave behavior is a peculiarity of certain experiments. Likewise the photon description of light requires its appropriate apparatus. The words 'wave' and 'particle' are carryovers from classical physics. In quantum mechanics, 'wave' and 'particle' are each associated with their own particular experiments.

But complementarity entails more than just wave-particle duality. Complementarity is the essence of Bohr's interpretation of quantum mechanics. It clearly distinguishes a classical experiment from a quantum one and it provides the necessary conceptual framework for the discussions that follow. We use 'complementarity' to mean the Bohr interpretation of quantum mechanics.

In this book complementarity is taken to include three non-classical characteristics of the quantum experiment:

1. Quantum experiments are entangled things.
2. The quantum experiment is not complete without a measurement result.
3. If we change the apparatus at any time prior to the measurement, the results obtained correspond to the experimental configuration in place at the moment particle is detected.

These are the defining features of quantum experiments. They are foreign to our classical mindset. In addition, Bohr insists that we use classical concepts and ordinary language in describing quantum events.

We now discuss each of these in detail because, taken together, they give us the answer to the question, "What is quantum mechanics about?"

4.2 The principle of non-separability (Entanglement)

Non-separability is at the heart of quantum complementarity. It is what we call entanglement. Bohr often referred to it as a 'wholeness' associated with the quantum experiment. It was Schrödinger who named it 'entanglement'. He believed it to be of fundamental importance. In his words,

> *Entanglement is not one, but rather, the*
> *characteristic trait of quantum mechanics,*
> *the one that enforces its entire departure*
> *from classical lines of thought.*

Quantum mechanics provides a description of entangled quantum experiments. It does not describe the behavior of individual things. This is quite different from classical physics where separability is the norm and each thing can be described individually. In the quantum experiment, where the finite value of Planck's constant cannot be ignored, entanglement is the rule. The different parts of the quantum apparatus lose their individual identities and the experiment is no longer a collection of its constituent parts. Classical physics and its principle of separability are no longer valid.

The real world of Einstein is separable into its constituent parts. The quantum world is not.

The entangled experiment is the essential element of quantum mechanics. It is the theme of everything that follows in this book.

> *The entire quantum experiment including all its parts, the particle source, the preparation apparatus, the particle, the measuring device, and the measurement result are all a single entity.*

In spite of what we see in the lab, quantum mechanics does not admit to the independent existence of the different parts of the experimental apparatus. Quantum experiments behave as if there are no separate individual pieces. Instead, they are all entangled into a 'whole' thing. They have a collective existence, unlike anything we have ever seen.

Bohr insists that there are no ordinary particles in a quantum experiment done with particles. Instead, the particle is lost, along with all the other parts, in the entangled whole. We tried to show this when discussing Schrödinger's cat, where we stressed that a quantum cat-in-a-box is quite different from an ordinary cat who happens to be in a box.

Bohr does not say there is no quantum reality. But he does say there are no independent realities of any separate parts to be considered. Individual particles and other individual devices no longer exist. If we think otherwise, then our calculations give the wrong answers. The only quantum reality is the entangled whole experiment.

Most importantly, the measurement device is an indispensable part of that experimental whole not to be neglected. It, too, has no separate existence. In particular, it is no longer independent of the particle being observed. They are now inseparable.

The separation between the observed particle and its measuring apparatus seen in the lab is an illusion. There is no such spatial separation in the quantum description.

Separability is one of the hallmarks of classical physics. It is a consequence of the fact that real things exist in space-time. Separability means that all real things are spatially separated from other real things, each with its own location in space. In particular, each real thing has its own physical attributes free and independent of any measuring device. Separability allows us to investigate real objects individually in different experiments, one at a time if necessary.

But, now entanglement deprives us of that ability. Entanglement denies a space-time existence for quantum entities. As a result, separability and locality, the pillars of classical theory, are abolished.

We pay attention only to the entangled whole where individual particles no longer exist. We are not describing a solitary particle in space-time with its own attributes as is done in classical physics. Rather, the physical attribute being measured has a value only in the context of the entangled experiment.

Yet, we do quantum experiments with particles. Our intuition and common sense tell us that a particle is emitted from a source, traverses the apparatus and later ends up in a particle detector. After all, that is how we designed the apparatus to function. But, the concepts of space, time, and causality have no place in quantum mechanics. The classical description given above no longer applies.

And, most troubling to many is that with quantum entanglement there is no underlying objective reality that makes the experiment work. There is no particle that undergoes a sequence of events while interacting with the various parts of the apparatus on its way to the detector. To do so requires that the particle and apparatus be separate things. They are not! We must not try to visualize how the particle gets from source to detector, even though we have pictures of the apparatus and we draw schematic diagrams to show how things are connected. Such imagery, a powerful asset in classical physics, is no longer available to us. There are no moving pictures of particles interacting with various devices as it moves through the apparatus. In fact, we have no idea how the particle is affected, if at all, by various devices in the apparatus.

This, of course, presents a serious problem for most of us. We cannot comprehend such a thing. Nothing in our experience has prepared us for such an alien concept. We know the experimental parts are all real things. After all, we assembled the apparatus from classical devices. We took the individual parts out of boxes or fabricated them ourselves in the lab. We can see them and touch them. There is nothing unusual about the pieces that we connected together. Each part exists at a specific location in the apparatus. It is there for a reason. We know how the pieces work individually and we know how they work together. That is the classical tradition.

Yet, when we turn on the switch the separate parts no longer function as expected, in a classical way obeying classical laws.

This does not mean, however, that atomic particles never behave like classical objects. Sometimes they do. It depends on the type of experiment done. For example, there is nothing out of the ordinary about the motion of electrons in a cathode ray tube used in old fashioned television sets. There, the electrons have trajectories determined by the laws of Maxwell and Newton. We know how the cathode ray tube works. Nothing unusual there!

Quantum mechanics, on the other hand, does not tell us how the entangled experiment works. In particular, it does not reveal anything about particle behavior before the measurement is made.

As John Archibald Wheeler has said in discussing a photon experiment,

> "---------we have no right to say what the photon is doing in all its long course from point of entry to point of detection".

We cannot follow 'a photon moving through the apparatus' because there is no individual 'photon moving through the apparatus'. Such a moving image requires a photon separate from the rest of the apparatus, a photon with a trajectory in 3-space. There is none of that in the quantum description. With entanglement in place, there is neither an ordinary particle nor a space for the particle to move through.

Only the whole experiment has any objective meaning.

We must keep reminding ourselves that quantum mechanics is about the results and the statistical distribution of those results obtained in the entangled experiment. Considering only those experimental facts is the only way to avoid the confusion that comes from speculating about what is going on during the experiment.

Quantum theory does not describe the inner workings of the entangled experiment, which yields only results and probabilities. According to Aage Petersen, when asked about any underlying reality, Bohr would answer,

> "There is no quantum world. There is only an abstract quantum physical description."

We must accept that a quantum particle is not an individual thing, the kind of thing we are used to. It is defined only as part of the entire apparatus. It is not defined in space-time. It does not have any dynamic properties, such as momentum, that are independent of the rest of the apparatus. Remember, Schrödinger's cat-in-a-box is not an ordinary cat. Rather, it is an entangled kind of thing that is unimaginable to us. So it is with any quantum object.

For example, there is enough evidence to support the claim that the hydrogen atom consists of a proton and an electron, just as Bohr speculated in 1913. But, the atom, a quantum thing, does not consist of two ordinary charged particles, real things interacting in a classical way under the influence of the Coulomb force. As we have shown, there is no such thing in the real world. The electron and proton do not exist separately in the hydrogen atom! It is physically impossible.

There is no electron and there is no proton in a hydrogen atom and we should not be talking as if there were. Rather, they now comprise an inseparable thing, entangled in a purely quantum way. The only objective reality is the whole hydrogen atom, the entangled thing. It is wrong to say, "There is an electron in the hydrogen atom." But, unfortunately, we do it all the time. We must understand that an atomic electron is not an ordinary electron.

Nevertheless, a quick look at images of atoms found in textbooks and on the internet invariably show individual electrons in a classical orbit about the atomic nucleus, just as Bohr imagined. We are unable to rid ourselves of such an erroneous visual representation even after decades of knowing it is wrong.

At least, there is now general agreement that classical physics doesn't work in the atomic domain. The entangled experiment is obviously not a classical object. And if we insist that the parts are real separate things obeying the laws of classical physics, then our calculated predictions are erroneous. We will see examples of this in the final chapters of this book.

4.3 A quantum experiment requires a result

Bohr was the first to recognize that the experiment is not finished without a measurement result. The measurement process, which ends with particle detection, is an irreversible event that gives closure to the experiment.

Closure occurs at the instant the particle is detected. The experiment is over and done when this occurs. Paraphrasing Wheeler,

> *"No quantum experiment is an experiment until there is an experimental result".*

(Actually, Wheeler said, "No elementary phenomenon is a phenomenon until it is a registered phenomenon.") Prior to measurement, that is before closure, the experiment is undefined. Quantum mechanics does not describe anything going on before closure.

There is no experiment to discuss until we have the measured value. For example, a quantum momentum experiment yields a measured value for momentum, a value that depends on the entire experimental arrangement in place at the instant the particle is detected. A quantum electron, for instance, does not have its own momentum the way the moon does. The electron has no momentum until there is a measured value, which depends on the conditions under which the measurement is made. That measured value is not a pre-existing value that the electron had all along. From here on, it is understood that 'value' always means 'measured value'.

Further, Asher Peres has warned us thus:

> *"Unperformed experiments have no results".*

As trivial as it sounds, it is, nevertheless, important to understand that quantum mechanics does not describe unperformed experiments, or unobserved atoms, for that matter. We must finish the experiment and obtain a measured value.

There is no objective reality that is always 'out there'. Rather, in quantum mechanics *an unobserved reality is meaningless.*

Useful information about a done experiment is obtained only from that particular experiment. We must not use data inferred from other mutually exclusive experiments that could be done at some other time. This is done all the time in classical physics, but is unacceptable in quantum mechanics where complementary attributes do not refer to the same experiment.

Suppose we do have two mutually exclusive experiments. We do the first experiment and later, we do the second one. There are no values from the earlier, but now unperformed, experiment that can be applied to the second experiment. Knowing the values obtained in the earlier experiment does not mean that those values are still true when the second experiment is done. Only the values obtained in the second experiment can be applied to that experiment.

In quantum mechanics, mutually exclusive means just what it says; doing one of two mutually exclusive experiments precludes doing the other. Mutually exclusive experiments are not part of the same quantum whole and their results cannot be considered in an inclusive way. They must be considered separately. We can discuss the results of the first experiment or we can discuss the second experiment, but never together. There is no single picture that embraces the results of mutually exclusive experiments.

Combining the results from two mutually incompatible experiments does not enhance our knowledge of either experiment. In fact, such counterfactual inference leads only to contradictions and confusion. We disregard this admonition at our own peril!

Complementarity and everything it engenders requires a radical revision of classical doctrine, which Einstein refused to accept. He said, "I refuse to believe that the moon does not exist when we don't observe it."

4.4 Which preparation procedure? Delayed choice

If we change the apparatus at any time during the experiment prior to particle detection, the results obtained correspond to the experimental configuration in place at the moment the experiment is closed. As expressed by Bohr,

> "-----*it can make no difference----------whether our plans ------- are fixed before hand or whether we prefer to postpone the completion of our planning until a later moment when the particle is already on its way from one instrument to another.*"

It is the final configuration that becomes the entangled whole. Even if we remove, or insert, a device long after the particle has supposedly moved beyond the location of that device, the obtained results are the same as if we had made the change before the particle began its journey. When we do make a change the new whole configuration replaces the old one instantaneously. So called delayed choice experiments confirm such baffling behavior.

In such experiments, we must assume that the particle passes through the entire apparatus in a classical manner from source to detector in order to determine where the particle is at every moment. But, again, we must remind ourselves that this is not classical physics. It doesn't matter where we think the particle might be when such a device change is made. A quantum particle does not have a space-time description. It isn't anywhere at the instant we make the change!

This doesn't make any sense to the classical physicist who insists that the particle has to be somewhere in the apparatus at every moment. He insists that it obey the classical rules. If the particle has already passed through a particular device, then removing that device after the fact should have no effect on the outcome of the experiment, since the particle would have already interacted with that device before its removal.

But entanglement replaces separability and locality. There is no particle moving through the apparatus, interacting with each device as it goes. That would be classical physics. Such a classical-like sequence of events does not occur in quantum experiments. The only question for us is, "At closure, is the device a part of the entire experimental apparatus or is it not?" In the quantum description, there is no 'before' particle detection.

We repeat, for emphasis, the most fundamental characteristic of the quantum experiment is the entanglement of all its parts and the subsequent loss of the individual particle concept. Quantum theory describes the results of the whole experiment. It does not describe a solitary particle. Nor is there any space-time description of the process that takes the particle from source to detector.

It has been confirmed in numerous experiments that the results of delayed choice experiments always correspond to the experimental configuration in place at the time of closure.

4.5 The use of classical concepts

As you can see from our presentation up to this point, it is not possible to do away with classical concepts all together when discussing quantum experiments. We use them all the time.

We still do experiments to measure position and momentum, for example. But, in so doing we lose the objectivity of the classical description.

Nevertheless, Bohr insisted that we describe quantum events in classical terms. He said,

> "---------- *the unambiguous interpretation of any measurement must be essentially framed in terms of the classical physical theories and we may say that in this sense the language of Newton and Maxwell will remain the language of physicists for all time. Otherwise, we will be unable ------ to tell others what we have done and what we have learned.*"

For example, the classical momentum does not require a measurement to have a value. A moving ordinary particle always has a definite momentum, independent of any measuring instrument. Not so in a quantum experiment! Momentum is still a measurable quantity, but it is no longer a property of the particle alone, and it is not simply mass times velocity. We cannot confirm the momentum value by repeating the same quantum experiment, since we usually get different results.

Even though we say 'momentum', this is not the same as the classical 'momentum'. It is defined differently. Generally, there is no answer to questions such as, "What is the particle's momentum?" We can only say that we have a measured value of momentum obtained in a given experiment.

Likewise, position is a classical concept. All classical particles have a location in space at every instant. But, quantum particles do not. An entangled quantum particle has only a measured value for position obtained in a particular experiment, just as it is for momentum.

We must learn new definitions for old concepts. We will do some of that in the next chapter. But, the greater difficulty is the continued use of classical words such as particle, experiment, device, apparatus, etc that imply a separability that no longer exists. This is the source of much confusion.

At the atomic level, our preconceived ideas about how things are supposed to work fail us.

4.6 The use of ordinary language

The only thing that is ordinary in quantum mechanics is the common language used, at Bohr's insistence, to describe quantum events. The experimental apparatus and the experimental result are described in ordinary language, the everyday language used to describe the classical world we see around us. Thus, we have an insoluble dilemma – we describe quantum phenomena with the language used to describe real classical events. The words we use in trying to describe quantum events originate in our everyday experience. They make it sound as if quantum things are real, as if they are no different from what we see around us.

The language used to describe the quantum world floods our minds with mental images that have no basis in fact. There is no visual representation of quantum events. Yet, we cannot help but imagine a smooth running atomic universe, a tiny version of the physical universe we observe around us. For example, it is easy to imagine an electron moving in a circular orbit about the proton in a hydrogen atom. Though always present, such images are wrong.

Take the word 'particle', for instance. This book is an attempt to describe particles on the atomic level. We use the word 'particle' all the time, but it does not depict a quantum particle correctly. It conjures up all sorts of images reminiscent of the real particles we are familiar with. Unfortunately, those images are false.

Here, 'particle' is the descriptor of choice for photons, electrons, atoms, and other like objects that are seen localized on detection screens.

But the word 'particle' implies a localized object, separate from everything else, moving through space independent of the measuring apparatus used to observe it.

This creates an unavoidable misrepresentation of what we are trying to describe. The language used is inadequate, at best, to correctly describe the entangled thing, what we call a 'quantum particle'.

Atomic events do not occur in the way we imagine them to. The true nature of the atomic world is so far removed from our everyday experience, that it is impossible to describe a quantum event with the language tools available to us. But, we have no choice. We must use our common language to communicate the results of our experiments to others. There is no other way.

The use of ordinary language that is part of our classical tradition constantly undermines the discussion. Unfortunately, this cannot be avoided.

4.7 The Stern-Gerlach Experiment

We will take the Stern-Gerlach experiment to be our quintessential quantum experiment. It is a real world version of the Schrödinger cat experiment where we use a magnet to measure the spin $1/2$, a purely quantum kind of angular momentum that has no counterpart in classical physics. Unbeknownst to Stern and Gerlach when they did the original experiment in 1922, the spin $1/2$ is a two-state observable and an experiment designed to measure it has only two possible outcomes.

Otto Stern and Walther Gerlach had intended to measure the magnetic dipole moment of the silver atom with the experimental apparatus shown in Figure 4.1. The original experiment used an oven with a hole in it as the source of electrically neutral silver atoms, which were then deflected as they moved through a non-uniform magnetic field. When in a magnetic field, atoms that possess a magnetic dipole moment behave like tiny bar magnets. We will first describe this experiment in classical terms because the classical physicist knows how bar magnets behave in a magnetic field.

Bar magnets are characterized by magnetic poles at each end. A bar magnet moving through a uniform magnetic field, has equal and opposite forces exerted on its opposite poles. Thus, the net force on each atom will be zero and, according to Newton's first law, each atom will continue to move in a straight line. There is no deflection of a magnetic dipole when passing through a homogeneous magnetic field.

Stern and Gerlach, being good classical physicists, knew this, so they used a magnet that had a non-uniform field. In such a field, the magnetic force exerted on one end of the atomic dipole will be different from the opposing force on the other end. Now, the net force is non-zero and the atoms are deflected away from their initial direction. The kind of magnet that provides the required non-uniform field is now known as a Stern-Gerlach magnet.

A classical calculation shows that the angle of deflection is determined by the orientation of the dipole moment in the Stern-Gerlach magnet. It was known that the silver atoms leave the oven with magnetic moments in all possible directions, so the atoms will be deflected at all angles $-\phi_{max} \leq \phi \leq \phi_{max}$. All scattering angles within this range are possible, depending on the direction of the dipole vector.

The apparatus of Figure 4.1 measures the z-component of the magnetic dipole vector. If the detector is a photographic plate, we expect to see a continuous smear in the z-direction, created by many atoms hitting the detector. This is all well understood by classical physicists. But, surprisingly, the Stern-Gerlach experiment doesn't work this way!

When the experiment was performed, the atomic beam was split in two, an upper beam at $\phi = +\phi_{max}$ and a lower beam at $\phi = -\phi_{max}$.

Instead of a continuous smear, only two dots were on the detection screen. All atoms in the upper beam had dipole moment vector pointing in the +z-direction and all atoms in the lower beam had dipole moment vector in the -z-direction. Apparently the z-component of the magnetic moment vector has only two possible values. If we count the number of atoms in each beam, we find that there is a 50% chance that the z-component is in the up direction and there is a 50% chance that it is in the down direction. But this doesn't make any sense, since the atomic beam entering the magnet should have magnetic moments randomly oriented in all possible directions. The results make it appear that all the incoming silver atoms have magnetic moments pointing only in the +z-direction or in the -z-direction.

Classical physics is unable to explain how the magnet does this. We have all possible directions of the magnetic moment vector going in, but only two directions coming out.

Here, the classical belief that we can explain what is happening as the atoms move from the oven through the apparatus fails us. There is no known physical process that could cause such unusual results. The atoms and the Stern-Gerlach magnet obviously do not interact in a known classical way. We must conclude that the laws of Newton and Maxwell fail us in this particular experiment.

The classical laws no longer apply. In particular, classical separability is no longer in effect. We are not observing separate atoms as they emerge from an oven. That is what we do in a classical experiment. But, this is not a classical experiment, and a silver atom in the Stern-Gerlach experiment is not an ordinary atom. (Remember the cat?) Quantum mechanics does not describe individual silver atoms and we are not measuring the z-component of the dipole moment vector as it enters the magnet.

Rather, a quantum calculation gives the z-component obtained in conjunction with the entire Stern-Gerlach apparatus. Now, the atom and magnet, along with everything else, is an entangled whole. There are no individual silver atoms separate from the measuring apparatus.

We should not be talking about atoms moving in a magnetic field, as the classical physicist would, since, obviously, this is not what is happening. We should be discussing the entangled whole experiment at the moment an atom impacts the detection screen.

THE STERN-GERLACH EXPERIMENT

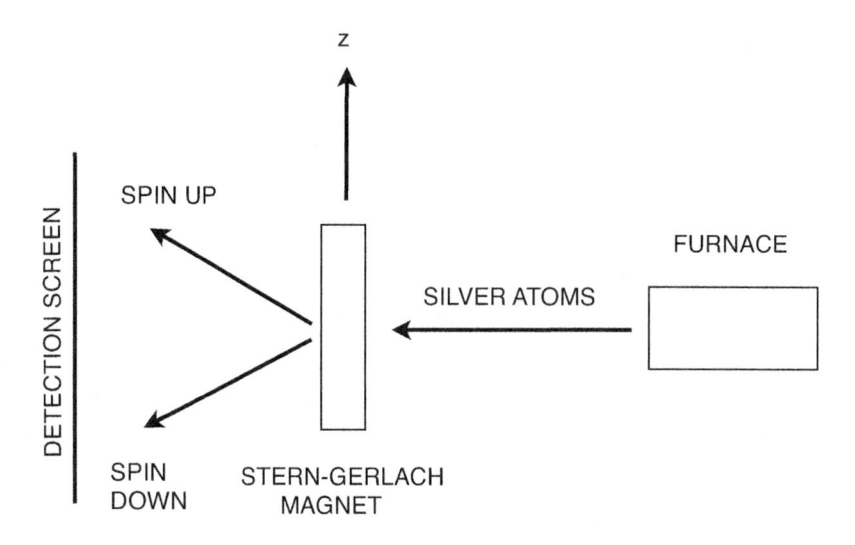

Figure 4.1. The Stern-Gerlach experiment measures the z-component of the magnetic dipole moment of neutral silver atoms. Classically, the deflection of an atom is determined by the orientation of its magnetic dipole moment. So, we expect to see all possible angles of deflection. Yet the atomic beam is split in two. All atoms are deflected only at an 'up' angle or at a 'down' angle. There is no classical explanation for these results.

As mentioned earlier, there is no 'before'. We must not speculate on what is happening before the atom strikes the detector screen. That is not what quantum mechanics is about. Quantum mechanics has no explanation for what is going on between the atom and the magnet.

We emphasize that the atom and the value of its magnetic moment are not separate and independent from the rest of the apparatus, even though the language used makes it sound as if they were.

As strange as it seems, there are no silver atoms moving from the furnace through the apparatus on their way to the detector. And, there is no known physical interaction that would cause the beam to split in two. According to Bohr, we should not waste our time trying to understand the inner workings of this experiment, because there are no 'inner workings' of this experiment.

The splitting of the beam is a purely random event. There is no reason why one atom goes 'up' and another one goes 'down'. There is no classical explanation of how the Stern-Gerlach experiment works. And, according to Bohr, no explanation is necessary. It just is!

We now know the Stern-Gerlach experiment measures an observable called 'spin 1/2', a kind of angular momentum, that is an inherent static property of quantum particles more akin to mass and charge. Mass and charge are single valued attributes, while the spin measured here has two possible values. A beam of spin 1/2 particles is always split in two by a Stern-Gerlach apparatus.

In modern versions of the Stern-Gerlach experiment, the beam intensity is so low that at any instant, only one particle is ever present in the magnet. From here on, we do the Stern-Gerlach experiment, not with a beam of particles, but rather one particle at a time.

Sometimes the particle is deflected up and other times it is deflected down. We do not speculate on how this happens.

Such is not the business of quantum mechanics and we no longer worry about it. We concern ourselves only with the result, the spin value obtained each time we run the experiment.

We see a dot where the particle impacts the detection screen. Particles deflected upwards at angle $\phi = +\phi_{max}$ are called 'spin up' (not very original, but that is the accepted convention) and its measured spin value is $+\frac{\hbar}{2}$. For 'spin down' particles, the deflection angle is $\phi = -\phi_{max}$ and the measured value is $-\frac{\hbar}{2}$. Usually, we divide out the $\frac{\hbar}{2}$ and report the two spin values to be +1 and −1. We will follow this common practice. Repeating the measurement many times, we observe that half of the time, a particle is observed to have 'up' spin while the other half of the time, it has 'down' spin. The probability of getting the value +1 is $P(+1) = .5$ and the probability of getting the value −1 is $P(-1) = .5$.

That's it! We are done with the quantum description of the Stern-Gerlach experiment. It has all the characteristics that we associate with a quantum experiment. It yields the two possible spin values and the probability of getting each value when we make a measurement. That is all there is. As we shall see, this is just what the formalism of quantum mechanics provides.

We still have no idea how the Stern-Gerlach experiment works. The splitting of the beam is simply a characteristic of the Stern-Gerlach experiment done with spin 1/2 particles. There is nothing more to say!

4.8 Spin vectors and a common trig identity

A vector is a concept from classical physics that is carried forward into quantum theory. Many physical quantities are defined as vectors in 3-space, position and momentum, among them. Classically, these vector quantities require three components for their complete specification. For example, we need three numbers (components) to determine the position of an object in three dimensional space. Can we say the same thing about the spin vector? Do spin vectors exist in 3-space as ordinary vectors do? They do not, and in this section we will demonstrate a difference between a spin 1/2 vector, a quantum thing, and an ordinary vector.

By definition, the component of a vector is its projection onto an axis at angle θ. It is a real number. The vector component is defined to be
$$V_\theta = V \cos\theta,$$
where θ is the angle that the vector makes with the axis in question, and V is its magnitude (length). A real vector has a component at every angle. The angle and its corresponding component are continuous variables.

Since $\cos 0° = 1$, $\cos 120° = -.5$, and $\cos 240° = -.5$, the sum of those particular components is zero. We have a trig identity that is satisfied for all vectors:
$$V_{0°} + V_{120°} + V_{240°} = V - .5V - .5V = 0.$$
There is nothing special about this. It is simply a mathematical identity that follows from the definition of the vector component. It has nothing to do with physics, but we do expect that all physical quantities that are vectors will obey it. At least in the classical way of doing things, the components of any vector quantity at those given angles will add up to zero.

The components are intrinsic properties of a vector. Every component has a definite value. When the value of a component is needed, we can always calculate it or we can measure it. The obtained value 'belongs' to that vector and is valid in all future calculations, not only in the identity given here.

Knowing the magnitude and direction of a vector, we can easily calculate the value of a component in any direction using the above definition. Or, you can draw the vector as an arrow on graph paper and measure the lengths of its projection on the specified axes, and then use the measured lengths, instead of the calculated values in the trig identity. The three measured values so obtained will work just fine, and we can use them where required. Either way, the calculated values or the measured ones will satisfy the given identity.

We now show that the spin vector, like other quantum things, is not ordinary.

As with all quantum attributes, the only spin value is its measured value. The only way to test the identity is with measured components of the spin 1/2 vector obtained in the Stern-Gerlach experiment. Such an experiment measures the component S_θ of the spin vector with the Stern-Gerlach magnet in the θ direction. We can change the orientation of the magnet at will to measure any component desired, but we can do only one measurement at a time.

An ordinary vector has only one component in the θ direction. Like other classical attributes, it has a definite value. The spin vector, on the other hand, has two possible values for the spin component in a given direction. The Stern-Gerlach experiment demonstrates that a component of a spin 1/2 vector does not have a definite value. Regardless of the magnet orientation, we always obtain the spin value $S_\theta = +1$, or we obtain the value $S_\theta = -1$.

In terms of the spin 1/2 components the identity to be tested is $S_{0°} + S_{120°} + S_{240°} = 0$. We measure the spin at $\theta = 0°$, then measure it at $\theta = 120°$, and finally at $\theta = 240°$.

Each experiment gives the measured value, either +1 or −1, to be used later in the identity. We repeat the same sequence of measurements and again sum the obtained spin values. Repeating many times we get all possible combinations of values to be used in the identity. These are shown in Table 4.1.

Classically, we expect the sum to be always equal to zero. Yet the three measured spin components never add up to zero. The identity is violated by each combination. It is impossible to satisfy the identity when the only allowed spin values are +1 and −1. This would never happen with vector quantities in classical physics.

The truth is, there is nothing wrong with the identity, which is actually a classical construct. If an experiment is done with classical vectors, the use of measured values is not a problem.

Classical physicists would use values obtained from three mutually exclusive experiments without apprehension and the identity is always satisfied. This counterfactual approach works just fine for ordinary vectors where $V_\theta = V\cos\theta$. However, in the Stern-Gerlach experiment, $S_\theta \neq S\cos\theta$. The spin components are not continuous variables.

However, there is a more fundamental problem here. Actually, it is impossible for us to test this inequality with component values of a spin 1/2 vector!

As in the classical approach, measuring different components requires mutually exclusive experiments, one for each component. It is impossible to know the three components in the same experiment. This is not a problem in classical physics. But in the quantum description, when we measure $S_{0°}$, it is the only spin component available to us. There are no values for $S_{120°}$ and $S_{240°}$ that can be used in the identity. Likewise for the other components. The spin vector is not an ordinary vector, even though the word 'vector' makes us think that it is.

ALL POSSIBLE COMBINATIONS OF SPIN VALUES USED IN THE IDENTITY

$S_{0°}$	$S_{120°}$	$S_{240°}$	$S_{0°} + S_{120°} + S_{240°}$
+1	+1	+1	+3
+1	+1	−1	+1
+1	−1	+1	+1
+1	−1	−1	−1
−1	+1	+1	+1
−1	+1	−1	−1
−1	−1	+1	−1
−1	−1	−1	−3

Table 4.1. All possible combinations of the results from three different Stern-Gerlach experiments are shown. We measure the spin component at 0°, then at 120°, and finally at 240°. We then add the values obtained. As shown, $S_{0°} + S_{120°} + S_{240°} \neq 0$. The vector identity is violated for all possible combinations. We expect the components of an ordinary (real) vector to satisfy this trig identity, but these measured components of the spin vector do not.

We must remember, again, that the only value is the measured value. The spin components, being quantum attributes, are properties of the entire entangled Stern-Gerlach apparatus, not of the particle alone. If we reorient the magnet, the old values no longer apply to the new experiment.

The spin vector has a component at $\theta = 120°$ only in the experiment where the magnet is set at $\theta = 120°$.

Even though we always get the value +1 or the value −1 whenever we measure the spin at 120°, we cannot assume that those same values are still valid when we measure the spin at other angles.

And, therein lies the difficulty. In order to test the trig identity we must have values for all three required components. We only have one! And, unfortunately, there is no single experiment that gives simultaneous values for all three of them. The spin 1/2 vector has only one component, the measured one.

In the quantum world of spin vectors we can always measure either $S_{0°}$, or $S_{120°}$, or $S_{240°}$, but, only one at a time. It is always possible to determine the value of each component separately, just so long as we don't use any two or more of those measured values in the same equation. The spin components are incompatible with one another. They are complementary attributes.

The experiments required to determine each component cannot be done simultaneously. The magnet can have only one orientation at a time and, consequently, the spin vector has only one component. When we measure the spin component at $\theta = 0°$, the experiments for $S_{120°}$ and $S_{240°}$ are not being done.

As we now know, such unperformed experiments have no results. 'No results' means just what it says: There are no values that can be used in our analysis. The only spin component is the measured component. There is never any value associated with an experiment not done.

It is not correct for us to say that that the spin 1/2 vector does not satisfy the given identity. In truth, there is no way for us to test the identity with spin 1/2 vectors. The results shown in Table 4.1 are incorrect. The correct values obtained in each Stern-Gerlach experiment are shown in Table 4.2.

SPIN VALUES OBTAINED IN THE THREE SPIN EXPERIMENTS

MAGNET SETTING	$S_{0°}$	$S_{120°}$	$S_{240°}$	$S_{0°} + S_{120°} + S_{240°}$
$\theta = 0°$	+1 −1			UNDEFINED
$\theta = 120°$		+1 −1		UNDEFINED
$\theta = 240°$			+1 −1	UNDEFINED

Table 4.2. The actual spin values for three mutually exclusive spin experiments. We have results only for the experiment actually done. There are no spin components from unperformed experiments. The trig identity requires three spin components. We can provide only one of them, making it impossible to test the identity.

We conclude that we cannot establish the validity of the trig identity with results taken from three mutually exclusive Stern-Gerlach experiments! We will need real experiments and components of real vectors in order to do that. We will continue to refer to the spin 1/2 'vector' and we will discuss experiments in which we measure its 'component' in a particular direction. But, we must accept the fact that the quantum spin vector is not an ordinary vector. It does not have a direction in 3-space.

Again, we have a word, 'vector', a classical thing, being used to describe a quantum entity. The ensuing erroneous approach to test the identity is a consequence of using classical concepts to describe a quantum thing. Quantum vectors, like other quantum things, defy a classical description and we should not expect them to satisfy classical identities.

5

THE POSTULATES OF QUANTUM MECHANICS
A set of instructions for doing quantum calculations

These symbols themselves, as is indicated already by the use of imaginary numbers, are not susceptible to pictorial interpretation; and are only to be regarded as expressing the probabilities for the occurrence of individual events observable under well-defined experimental conditions.

Niels Bohr

5.1   General

Quantum mechanics is actually a rather meager theory. As we have said, it only provides two bits of information about quantum events, the possible results of a measurement and the statistical distribution of those results. Taking quantum mechanics at face value, that's all there is!

The quantum postulates given below provide a set of instructions for obtaining that information for a given experiment.

The postulates will give my non-physicist readers a sense of how physicists actually 'do' quantum mechanics. The mathematical formalism is straightforward, but, as is often the case, the devil is in the details. A course in linear algebra provides the budding physicist with the required mathematics.

By the mid 1920s, Bohr's original theory of atomic electrons jumping from one stationary orbit to another, had been rejected, but a comprehensive theory had yet to be found. The intense intellectual struggle to find the mathematical structure for the physics of atomic systems began with Louis deBroglie's PhD thesis in 1924. From that moment on, it took about four years to figure out the sought after mathematical formalism, a remarkably quick period of discovery in the annals of science. There were many contributors, but Bohr himself did not contribute directly to the developing formalism. There is no 'Bohr Equation' or 'Bohr Postulate' to help us make the required calculations. Nevertheless, he was the recognized master of it all. He prodded and harangued Heisenberg and Schrodinger, among others, until they got it right. If you wanted to be a part of the revolution, you went to Copenhagen.

The postulates given here do not provide for a description of particles moving in space-time as is done in classical physics. Nor is there any reference to any mechanism that is responsible for the obtained result. All the details given to us in the classical description are nowhere to be found. In short, there is no reference to any inner workings of the quantum experiment. Rather, the formalism admits to only the purely statistical nature of experimental results.

And there in the midst of the new mathematics, some of it unfamiliar to many physicists of that era, was Planck's constant. Here in the quantum domain, the finite size of Planck's constant cannot be neglected. Now, whenever we see an equation containing h or $\hbar$, we know it originated in quantum mechanics.

We must get mathematical at this point. Since there is no pictorial representation for atomic systems and since the language used is faulty, only the mathematical description of quantum events is true and accurate. We can believe the math. Everything else is suspect! If we are to attempt an understanding of quantum mechanics, we cannot ignore its mathematical underpinnings. And with these postulates we surely leave the classical world of real things.

I want you to know something about the vocabulary of quantum mechanics, much of which has its source in the mathematical formalism. If you do much reading about these things, you will often see the terms introduced here and I would like you to have some idea of their meanings.

The notation used was invented specifically for quantum mechanics by Paul A. M. Dirac, one of the founding fathers. It is truly the language of quantum mechanics and if you are a physics major, you may already be familiar with it. But, if you are uncomfortable with it, then you can skip over those sections that bother you the most. But I hope you will give it a try.

5.2 The postulates of quantum mechanics

Postulate 1. Definition of observable
The observables of quantum mechanics are Hermitian operators.

In quantum mechanics, many of the physical quantities we are familiar with from classical physics are replaced by symbolic operators, which are very different from their classical definitions.

SOME PHYSICAL PROPERTIES

	CLASSICAL PHYSICS	QUANTUM MECHANICS
POSITION	x	$\hat{x} = x$
MOMENTUM	$p_x = mv_x$	$\hat{p}_x = -i\hbar \dfrac{\partial}{\partial x}$
ENERGY	$E = \dfrac{p_x^2}{2m} + V(x)$	$\hat{H} = -\dfrac{\hbar^2}{2m}\dfrac{\partial^2}{\partial x^2} + V(x)$
SPIN 1/2		$\hat{s}_z = \begin{bmatrix} 1 & 0 \\ 0 & -1 \end{bmatrix}$

Table 5.1. Some properties of a particle confined to the x-axis as defined in classical physics compared to their definitions in quantum mechanics. Here, E is the total energy and $\hat{H}$ is the corresponding total energy operator, called the Hamiltonian. We use the ^ to indicate the symbol is an operator. Likewise, the momentum operator $\hat{p}_x$ is unlike the classical momentum $p_x = mv_x$, where m is the particle mass and v_x is the x-component of its velocity.

Quantum momentum, for example, is a differential operator, while the classical momentum is mass times velocity, an ordinary vector with real components. The mathematical differences between classical and quantum theory for some common observables are given in Table 5.1. We want you to see the obvious differences.

Position, momentum, energy, and spin, among others, are physical properties to be measured. In quantum mechanics, they are all Hermitian operators.

As we have said many times, a physical property has a definite value in classical physics. For example, a moving particle has a unique momentum value at every instant. It doesn't matter whether we have a measured value or not. We need not know its value.

But, now the momentum is an operator, a purely mathematical construct. The momentum doesn't 'belong' to anything and, in general, it doesn't have a definite value. Generally, there is more than one possible result when we measure it.

Some quantum observables have classical analogs and others do not. For example, the classical concepts of energy, position, and momentum are used in both the classical as well as the quantum descriptions. According to Bohr, this is as it should be - we describe quantum systems in terms of classical concepts. An exception is the spin, a kind of angular momentum not found in classical physics.

When the experiment in which an observable is measured has a finite number of discrete results, the observable is a Hermitian matrix operator. For example, we say that electron spin is a two-state observable, because there are only two possible values when spin 1/2 is measured. As shown in Table 5.1, the spin operator is a 2×2 matrix. As we have seen, the well being of Schrödinger's cat is also a two-state observable.

We emphasize that observables are not pre-existing properties possessed by the particle and, most important in quantum experiments, the obtained values do depend on the measuring apparatus. An electron has momentum, for example, only if we have a measured value in a specific experiment.

As stated earlier, the only 'value' is the 'measured value' and that measured value is a property of the entire experimental apparatus.

Postulate 2. Measurement of an observable
The only possible results of a measurement are the eigenvalues a_k of the measured observable $\hat{A}$.

Quantum mechanics does not predict the experimental outcome. But this postulate tells us how to obtain the list of all possible results when we make the measurement. That list is the eigenvalue spectrum of the observable being measured. For example, $+1$ and -1, the only possible results of a spin 1/2 measurement, are the eigenvalues of the spin 1/2 operator.

Postulate 3. The state vector
For a two-state observable $\hat{A}$, the quantum state is represented by the two dimensional state vector:
$$|\psi\rangle = \psi_1|a_1\rangle + \psi_2|a_2\rangle,$$
where the basis vectors $|a_1\rangle$ and $|a_2\rangle$ are eigenvectors of the measured observable $\hat{A}$ corresponding to the eigenvalues a_1 and a_2 respectively.

We see that the state vector is a linear superposition of eigenvectors. For a two-state observable, it has two components $\psi_1 = \langle a_1|\psi\rangle$ and $\psi_2 = \langle a_2|\psi\rangle$. The component $\langle a_k|\psi\rangle$ is a generalization of the scalar product, the so-called dot product known to us from ordinary vector algebra. Knowing the state vector means that we know all its components.

State vectors are only slightly different from the ordinary vectors in 3-space that we are familiar with. There are two main differences:

1. Quantum state vectors can have more than three dimensions, depending on the observable being measured.
2. Generally, the components of a state vector are complex numbers.

In other words, the state vector is defined in a complex linear vector space, not in the real 3-space where we do our experiments. As such, it does not move through the experimental apparatus and it does not interact with the measuring device.

We must continually remind ourselves that the state vector is not a real objective entity. Also, it is not part of the entangled whole.

The state vector so defined represents the entangled whole experiment in place at the moment the particle is detected. Any change in the experimental arrangement necessitates a different state vector.

The eigenvectors and eigenvalues of Hermitian operators are central to the mathematical formalism. They are obtained by solving the eigenvalue equation of the observable being measured. The postulates depend on two most important properties of Hermitian operators:

1) The eigenvalues of Hermitian operators are real numbers.
2) The eigenvectors of Hermitian operators are orthogonal:

$$\langle a_1 | a_2 \rangle = \langle a_2 | a_1 \rangle = 0.$$

The eigenvalues must be real because the measured values are real numbers. When we use Hermitian operators to predict the possible outcomes of an experiment, our calculation always gives real numbers for the measured values, as it must.

Postulate 4. The Born probability interpretation

The probability that a measurement of observable A yields the value a_k is

$$P_k = |\psi_k|^2 = |\langle a_k | \psi \rangle|^2,$$

where $|\psi\rangle$ is the state vector.

The quantum probability so obtained is an unavoidable consequence of the indeterminate nature of the state vector. Classical physics provides all the details about a classical event, while quantum mechanics provides only probabilities about the event. Quantum systems are inherently probabilistic and we describe a quantum event by giving the probability that the event will occur. If you are going to do quantum mechanics, then you will be solving eigenvalue equations and calculating probabilities. Probabilities are fundamental.

Once we have obtained the state vector, the Born postulate tells us what to do with it. The state vector written as a superposition of eigenvectors is a mathematical construct used to calculate probabilities. Nothing more! Anything beyond that is mere speculation.

Since the probability of getting the value a_k is $P_k = |\psi_k|^2$, the component $\psi_k = \langle a_k | \psi \rangle$ is called a probability amplitude.

As an example, the state vector for the Stern-Gerlach experiment is written as a superposition of spin eigenvectors:

$$|\psi\rangle = \psi_+|+\rangle + \psi_-|-\rangle$$

$$= \sqrt{\frac{1}{2}}|+\rangle + \sqrt{\frac{1}{2}}|-\rangle,$$

where the probability amplitudes are $\psi_+ = \sqrt{\frac{1}{2}}$ and $\psi_- = \sqrt{\frac{1}{2}}$.

We emphasize that the superposition state representing the

Stern-Gerlach experiment does not mean that the electron has spin values +1, and −1 at the same time. Rather, it allows us to calculate the probability for getting the value +1 and to calculate the probability for getting the value −1 when we do the Stern-Gerlach experiment with spin 1/2 particles.

Repeating the Stern-Gerlach experiment many times, the Born postulate predicts that we get the spin value +1 in half of the measurements and we get the spin value −1 in the other half. But we never get two different values in the same measurement. The orthogonality of the eigenvectors $\langle +|-\rangle = \langle -|+\rangle = 0$ prevents it.

When the state vector is the eigenvector $|+\rangle$, the probability of getting the spin value −1 is $|\langle -|+\rangle|^2 = 0$. We can never get −1 when the spin value is +1. Likewise, $|\langle +|-\rangle|^2 = 0$. The spin value can never be +1 when it is known to be −1. The spin is always found to have only one value, either +1, or −1, when it is measured. Likewise, the state vector for the Schrödinger's cat experiment is

$$|\psi\rangle = \sqrt{\frac{1}{2}}|alive\rangle + \sqrt{\frac{1}{2}}|dead\rangle.$$

There are two possible results for this experiment: the cat is alive or the cat is dead when we open the box. The probability of finding a live cat is the same as finding a dead cat when the box is opened:

$$P(alive) = P(dead) = \frac{1}{2}.$$

Again, the eigenvectors are orthogonal,

$$\langle dead|alive\rangle = \langle alive|dead\rangle = 0,$$

meaning we never find a dead cat that is also alive, or a live cat that is also dead.

In a superposition state, as is the general case, the measured observable is indeterminate. We do not expect to get the same value for the measured observable every time we do the same experiment.

Repeated measurements with the same experimental apparatus will generally have results different from one another. This is characteristic of quantum experiments. If we keep making the same measurement, we will eventually generate the entire eigenvalue spectrum of the observable being measured. The probability of occurrence for each eigenvalue result is calculated from the state vector using the Born postulate.

We stress that the state vector does not refer to an individual quantum particle. For example, an electron does not have its own unique state vector. Rather, its state vector represents the entire experimental arrangement in place at the moment that the electron is detected.

Postulate 5. Schrödinger's equation - stationary states
The quantum state function evolves in time according to Schrödinger's time-dependent equation

$$i\hbar \frac{\partial \Psi(x,t)}{\partial t} = \hat{H}\Psi(x,t).$$

Here, $\hat{H} = \frac{\hat{p}^2}{2m} + \hat{V}(x)$ is the total energy operator, called the Hamiltonian, $\frac{\hat{p}^2}{2m}$ is the kinetic energy operator, and $\hat{V}(x)$ is the potential energy operator.

The Schrödinger wave function $\Psi(x,t)$ is the probability amplitude for finding the particle at position x at time t. Some prefer to call it a probability wave. It is used to calculate the probability density $|\Psi(x,t)|^2$.

Schrödinger's equation is not a classical wave equation and $\Psi(x,t)$ is not an ordinary wave. A real wave function is a solution of the classical wave equation, but it is never a solution of the Schrödinger equation. Because of the imaginary unit, $i = \sqrt{-1}$, its solutions are necessarily complex functions.

Schrödinger's time dependent wave function does not describe a particle moving from one position to another. Nor does it represent a changing physical process. Where applicable, it allows us to determine the time evolution of the probability distribution of position values obtained in a given experiment. Nothing more.

Schrödinger initially wanted his wave function to represent real entities, but he couldn't make it work. As far as we know, Schrödinger waves, like state vectors defined earlier, are strictly mathematical symbols. They do not possess energy or momentum. They do not propagate in 3-space and they do not interact with measuring devices. Further, there is no vibrating medium for quantum wave functions. We must not think otherwise.

N. G. van Kampen has warned us thus:

"Whoever endows Ψ with more meaning than is needed for computing observable phenomena is responsible for the consequences........"

The wave function does not describe particle behavior. In particular, we emphasize that the wave function does not describe a particle that is smeared out all over a detection screen. The 'smearing' requires many different particles hitting the detection screen one particle at a time.

As we have said many times, the mathematical structure allows us only to calculate the possible results and the probability distribution of those results when we make measurements.

The Hamiltonian operator in Schrödinger's equation gives the total energy a special place in quantum theory. For an isolated system the Hamiltonian is independent of the time and the general solution of the Schrödinger equation is $\Psi(x,t) = \psi(x)e^{-i\omega t}$.

Here, $\psi(x)$ is an eigenfunction of the total energy operator H corresponding to the energy eigenvalue $E = \hbar\omega = hf$, which we recognize as Planck's hypothesis.

Energy eigenstates are important because probabilities do not evolve in time when the state function is an energy eigenfunction. Energy eigenfunctions are stationary state solutions of the Schrödinger equation.

For example, in Bohr's hydrogen atom the energy states that correspond to his non-radiating orbits are eigenstates of the Hamiltonian operator. Bohr had unknowingly discovered the stationary states for the hydrogen atom.

5.3 Quantum uncertainty

Because we cannot predict the outcome of a quantum measurement, we say that quantum mechanics is indeterminate, or 'uncertain'. The observable being measured does not have a definite value as in classical physics. The statistical distribution of the experimental results is a consequence of this inherent indeterminate nature of a quantum experiment.

We define the uncertainty in observable A, written ΔA, to be a measure of that intrinsic indeterminacy.

But, there is a special case where $\Delta A = 0$. In its eigenexperiment, the observable A does have a definite value a_1, say. There is only one possible result for an eigenexperiment, so it is deterministic, as in classical physics. Repeating the same eigenexperiment for observable A always gives the same result and its state vector is the eigenvector $|\psi\rangle = |a_1\rangle$

corresponding to the eigenvalue a_1. We know the value of observable A with certainty, meaning $\Delta A = 0$. This is as close to a classical experiment as we can get. The statistical distribution of the measured values obtained in an eigenexperiment are shown in Figure 5.1.

```
                    STATISTICAL DISTRIBUTION OF MEASURED
                       VALUES WITH ZERO UNCERTAINTY
PROBABILITY │
            │                    │
            │                    │
            │                    │
            └────────────────────┴──────────────────────→
                         MEASURED VALUES
```

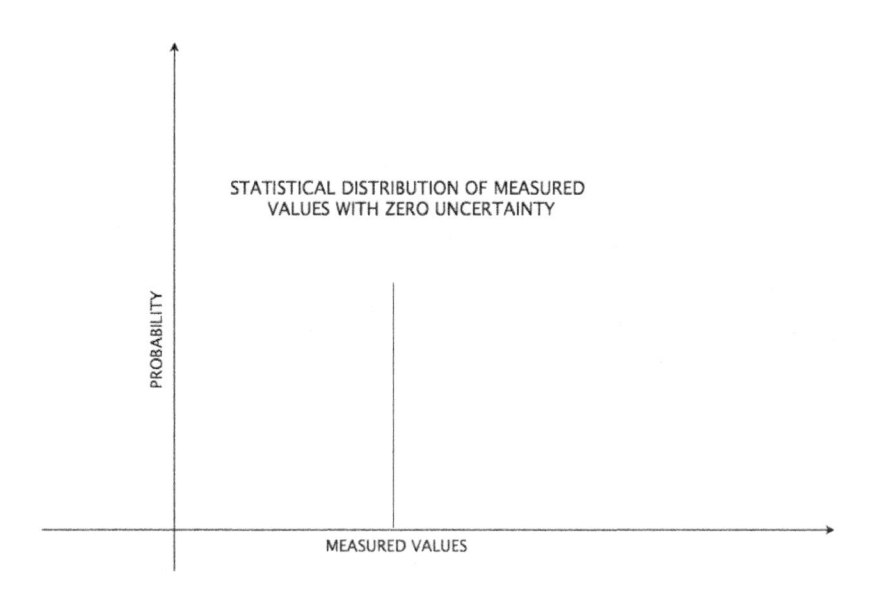

Figure 5.1. Here, we see the results of an eigenexperiment where we always get the same result in repeated measurements. In this case, the probability distribution is a 'spike' without any spread in the measured values. We know the measured value with certainty.

But, in general, quantum experiments are not eigenexperiments and they are not determinate. The typical quantum experiment generally has more than one possible result and there is always a non-zero uncertainty in the attribute being measured.

Although we cannot predict the outcome of such an experiment, the state vector does allow us to calculate the uncertainty ΔA along with the probability $P(a_k) = |\langle a_k | \psi \rangle|^2$ for each possible result.

Uncertainty, indeterminacy, probabilities, and superposition states are inherent characteristics of quantum mechanics. These are just different ways of saying the same thing. They are all manifestations of the indeterminate nature of quantum events.

Much to the chagrin of Einstein, $\Delta A \neq 0$ means that doing the same thing over and over again does give different results. But quantum mechanics does determine the statistical distribution of all the values obtained in repeated measurements. If the statistical distribution of the measured results is narrow, the uncertainty is small. But, when the statistical distribution is spread over a wider range of measured values (eigenvalues), then the uncertainty is greater.

What we call uncertainty, then, is a measure of the spread in the statistical distribution of the experimental results. Figure 5.2 shows a probability distribution for which the uncertainty is small, but non-zero. These results are from an almost eigenexperiment.

In Figure 5.3 the experiment obviously has a greater spread in the measured values, so the uncertainty is larger still.

It is important to understand that quantum probabilities are not the same as the probabilities of classical statistical analysis. In quantum experiments we do not obtain different values of the measured observable because of imperfect measuring devices or because of human ignorance.

Quantum uncertainty is a characteristic feature of quantum experiments, independent of the inaccuracies of real measuring devices.

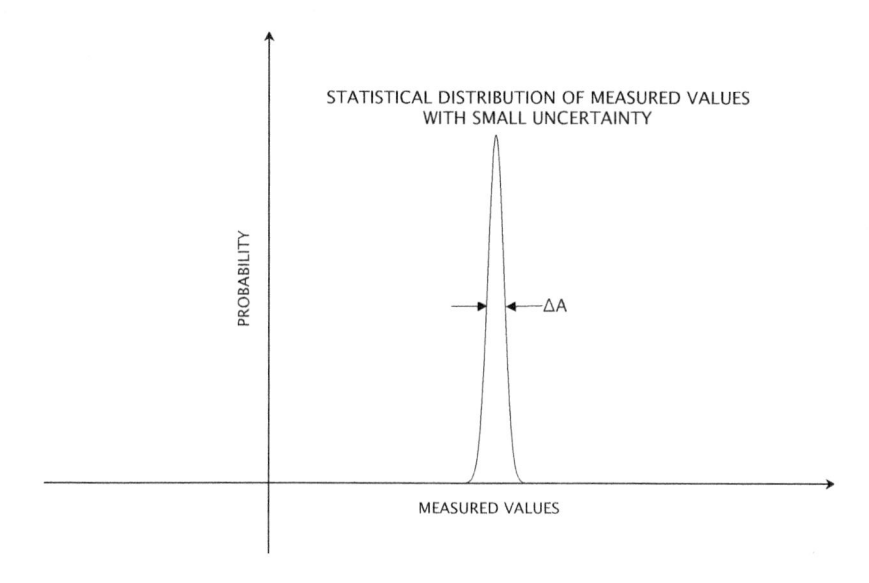

[Figure: Graph showing a narrow peak probability distribution centered on measured values, labeled "STATISTICAL DISTRIBUTION OF MEASURED VALUES WITH SMALL UNCERTAINTY", with width marked as ΔA. Y-axis: PROBABILITY, X-axis: MEASURED VALUES.]

Figure 5.2. Here, the measured observable has a continuous eigenvalue spectrum. The probability distribution has a narrow spread about some dominant value. The experiment is almost an eigenexperiment and the quantum uncertainty is small.

We repeat for emphasis - the quantum uncertainty has nothing to do with the errors induced by the limitations of the experimental apparatus. In fact, the discussion here, assumes perfect instruments, ignoring the experimental errors that necessarily occur in all human endeavors. Yet, even with perfectly run experiments, there is still an uncertainty in our results.

In classical physics, any spread in the measured results is due to errors in the measuring process. In the classical world, more accurate instruments yield a more narrow distribution of the measured values and perfect instruments always yield the same result with zero error.

Classical experiments are always determinate and any spread in the obtained results must be due to faulty instruments or human ignorance.

The statistical distribution from a classical experiment done with perfect instruments, then, is identical to Figure 5.1. But this is not what happens in a quantum experiment where the state vector is a superposition state. In a superposition state, perfect instruments do not make for zero uncertainty. We still get a spread in the measured values.

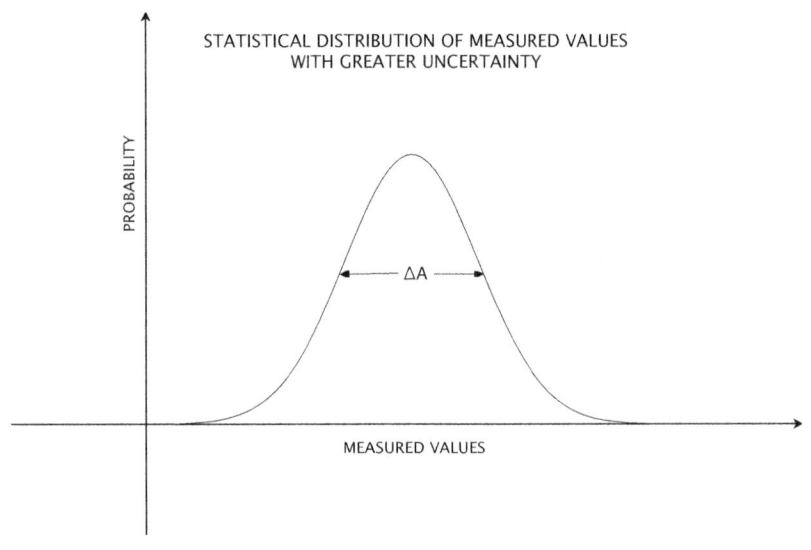

Figure 5.3. A greater spread in the statistical distribution of the measured values, as shown here, is indicative of a greater uncertainty. Most quantum experiments are not eigenexperiments. Such non-eigenexperiments have state vectors that are superpositions of the eigenvectors of the observable being measured. A superposition state always means a positive non-zero uncertainty.

It is important that we understand that quantum uncertainty and our use of probabilities has nothing to do with our inability to make perfect measurements. The point is this - we can make very precise measurements and still have a large quantum uncertainty in the obtained values. Because quantum events are naturally probabilistic, the uncertainty in our measured values cannot be avoided.

In summary, when we measure observable A,

a) The uncertainty is zero only when the state vector is an eigenvector of the observable measured. There is only one possible measurement result.
b) There is a positive non-zero uncertainty when the state vector is a superposition of eigenstates. There is more than one possible measurement result.

The important thing to remember is that it is the state vector $|\psi\rangle$, not imperfections in our measuring devices or lack of knowledge, that is responsible for the uncertainty.

5.4 The Heisenberg uncertainty principle

A classical particle has many attributes, all of which are well defined. Each might require its own experiment to determine its value, but each has a definite value, nonetheless. We can measure each attribute one at a time and the obtained values are all relevant with zero uncertainty. For example, a classical particle has definite values for both position and momentum at every instant. This is not so with quantum particles because of the inherent uncertainty in their measured results.

Generally, quantum observables do not commute; $AB \neq BA$. Observables that do not commute do not have common eigenvectors or common eigenexperiments. Since a quantum observable has a definite value only in an eigenexperiment, non-commuting observables cannot both have definite values in the same experiment.

If observable A is measured in its eigenexperiment, then it is well defined, the state vector is an eigenvector of A, and repeated measurements always yield the same eigenvalue. For such an experiment, $\Delta A = 0$. However, with the same eigenexperiment, repeated measurements of observable B do not give the same result and $\Delta B \neq 0$.

The superposition state defined above assumed a discrete set of eigenvalues and the state vector was a series expansion in terms of eigenvectors. But, that is not always the case. In particular, position and momentum are important observables that have continuous eigenvalue spectra. Then, the state vector becomes a state function, and instead of a series expansion in terms of eigenvectors, we have an integral expansion in terms of eigenfunctions.

When we measure position or momentum, then, we use a state function, often called a wave function, to make our probability calculations. The wave function is actually a probability amplitude. For example, the probability amplitude $\psi(x) = \langle x|\psi \rangle$ is the wave function for a position measuring experiment and $|\psi(x)|^2$ is the corresponding probability density for finding the particle at position x. Likewise $\psi(p) = \langle p|\psi \rangle$ is the state function when measuring the momentum in the same experiment.

Like other Hermitian operators, the position operator has orthogonal eigenfunctions. This means that a particle can never be found in more than one place at a time, in spite of what you may have read elsewhere. Quantum mechanics says that when we make a position measurement, there are an infinite number of possible locations where the particle might be found. But each measurement finds the particle at a single location. Repeating the experiment involves a different particle usually found at some other location. Knowing the position wave function, we can calculate the probability for finding a particle at each possible location.

In classical physics, the components of position and momentum do commute. For example, for a particle moving in the x direction, $p_x x - x p_x = 0$. As you know, this is true for all real numbers. In 1925 it was Werner Heisenberg who discovered that this simple expression is not true in quantum mechanics. Obviously, position and momentum are no longer real numbers.

We now know that position and momentum are non-commuting Hermitian operators with

$$p_x x - x p_x = -i\hbar.$$

Since position and momentum do not commute, there is no function that is an eigenfunction of both. It is mathematically impossible. Hence, there is no experiment that is an eigenexperiment of both position and momentum.

Although Heisenberg did not tell us how to calculate the uncertainty, most now define it to be the standard deviation (also called the root-mean-square deviation) known to us from classical probability theory. Using that definition, the product of the position uncertainty and the momentum uncertainty satisfies the inequality

$$\Delta x \Delta p_x \geq \frac{\hbar}{2}.$$

This is the celebrated Heisenberg uncertainty principle. It is a consequence of the fact that position and momentum are non-commuting observables.

The Heisenberg uncertainty principle says that there is no experiment for which both $\Delta x = 0$ and $\Delta p_x = 0$. Since a trajectory requires definite values for both position and momentum at every instant, the Heisenberg uncertainty principle is another way of saying that a quantum particle has no trajectory.

Suppose that we construct an experiment which is not an eigenexperiment for either position or momentum. When we make position measurements, the position state function is a superposition of position eigenfunctions. Repeating the experiment many times, we get a spread in the statistical distribution of position values. In such a superposition state, the position measurement has non-zero uncertainty $\Delta x > 0$.

Likewise, measuring the momentum with the same preparation apparatus, the momentum state function is also a superposition state and there is a corresponding momentum uncertainty $\Delta p_x > 0$.

There is a spread in the measured values of momentum as well as a spread in the position measurements. Position and momentum each have non-zero uncertainties. The actual values for Δx and Δp_x are obtained from the state function. And, in no case, will the product of the uncertainties ever be less than $\frac{\hbar}{2}$.

In this way, the Heisenberg Uncertainty Principle is built into the quantum state function.

It is true that we can design an experiment that is a position eigenexperiment for which there is no spread in the statistical distribution of position values. We will always find the particle at the same location, so that $\Delta x = 0$. But, if we should decide to measure the momentum in the position eigenexperiment, there is an infinite spread in the statistical distribution of momentum values; in order to satisfy Heisenberg's uncertainty principle, $\Delta p_x = \infty$. when $\Delta x = 0$. In such a case, we can see why it is meaningless to ask, "What is the particles momentum at position x?"

There are two common misconceptions about the Heisenberg uncertainty principle that should be addressed. First, it does not mean that you cannot measure position and momentum simultaneously. You can. In fact, the Heisenberg uncertainty principle requires that the same wave function, which refers to a given preparation apparatus, be used to calculate both uncertainties.

Sometimes, you can measure position and momentum in the same experiment, but, that experiment is not an eigenexperiment of either one. The position eigenexperiment and the momentum eigenexperiment are two different, mutually exclusive experiments that cannot be done at the same time. It is physically impossible. We can do one eigenexperiment or the other, but never both simultaneously.

The point is this: we can never have $\Delta p_x = 0$ and $\Delta x = 0$ in the same experiment. Position and momentum are complementary observables. In fact, Bohr believed the Heisenberg uncertainty principle to be a consequence of complementarity. He preferred to call it an 'indeterminacy relation'.

Second, the Heisenberg inequality does not place any restrictions on the accuracy of our measured values. Heisenberg's inequality is a relationship between uncertainties, between position uncertainty Δx and momentum uncertainty Δp_x. It is not a relationship between the measured value of the position x and the measured momentum value p_x.

The accuracy of a measured value is limited only by the accuracy of the measuring device. Perfect instruments yield exact values, yet there is is still a quantum uncertainty in the measured value. We can have very accurate values for both position and momentum in individual measurements. It is the uncertainties, not the values themselves, that are restricted by the Heisenberg uncertainty principle.

Generally, whenever we make quantum measurements we expect a spread in the statistical distribution of the measured values and there will be a corresponding non-zero uncertainty. Uncertainties are part of the quantum landscape. The measured values can be very accurate, but indeterminate, nonetheless.

Further, if we measure any two non-commuting observables in the same experiment, there will be a Heisenberg-like uncertainty relation between their respective uncertainties.

Observables that do commute have common eigenvectors and common eigenexperiments. Only then do both observables have definite values with zero uncertainties; $\Delta A = 0$ and $\Delta B = 0$ in their common eigenexperiment.

5.5 de Broglie's hypothesis

As an example of how we apply quantum mechanics, we consider an experiment where we measure the momentum of a free particle moving in the x-direction.

In order to make our quantum predictions, we need to know the Hermitian operator that represents the linear momentum for a free particle. Fortunately, the quantum operators commonly used are well known to us. The early pioneers of quantum theory discovered them all long ago. As shown in Table 5.1, the x-component of momentum is the differential operator

$$p = -i\hbar \frac{\partial}{\partial x}.$$

In classical physics there are no forces acting on a free particle. The particle has a constant momentum at every point on its trajectory. But, now if we want to know the quantum particle's momentum, we must measure it. As with other quantum attributes, we don't know the momentum value until we have a measured value. Generally, the measured value is not a property of the particle alone. It is characteristic of the entire measuring apparatus. We must do an experiment in which we actually measure the momentum. The experimental value so obtained will be one of the real eigenvalues of the momentum operator.

Solving the eigenvalue equation for the momentum operator, we get the eigenfunction e^{ikx} corresponding to the eigenvalue $p = \hbar k$. Here, $-\infty < x < \infty$, and $k = \frac{2\pi}{\lambda}$, from which we obtain deBroglie's relation $p = \frac{h}{\lambda}$.

Only if we do a momentum eigenexperiment, does a free particle have a definite momentum value with $\Delta p = 0$. The state function is then the eigenfunction e^{ikx} with definite wavelength $\lambda = \dfrac{h}{p_x}$. In the momentum eigenstate we know the momentum value p_x, as well as the corresponding deBroglie wavelength λ, with certainty.

But, most of the time the momentum state function is not an eigenfunction, the particle does not have a definite momentum value, and the corresponding state function does not have a definite wavelength. Rather, the state function is then a superposition of the momentum eigenfunctions and $\Delta p \neq 0$. Repeated measurements generate the entire spectrum of momentum eigenvalues. The momentum does not have a unique value, and, consequently, neither does its associated deBroglie wavelength.

We see, then, that deBroglie's hypothesis refers to momentum eigenstates of a free particle, where the momentum eigenvalue and its associated wavelength have definite values:

For a free particle of momentum p, the momentum eigenfunction has wavelength

$$\lambda = \frac{h}{p}.$$

Of all possible momentum state functions, only the eigenfunction e^{ikx} has unique wavelength λ corresponding to eigenvalue p.

The confusion concerning such 'wave-particle duality' is often a consequence of identifying the particle with its corresponding state function. Many are wont to say, "A particle of momentum p has wavelength $\lambda = h/p$." But, the deBroglie wavelength λ is a property of the momentum eigenfunction e^{ikx}, not of the particle.

It is important to note that the free particle is part of the entangled whole in the laboratory where we actually do the quantum experiment. But, its eigenfunction e^{ikx}, like other state functions, resides in a complex linear function space. It is an abstract mathematical thing without any attributes that can be seen in the laboratory. We cannot visualize it. Its complex nature renders it incapable of detection in the laboratory. Further, it is not part of the entangled experimental whole.

As we have seen elsewhere, it is Planck's constant that links the two contrasting properties, the particle momentum and the eigenfunction wavelength.

5.6 The single slit scattering experiment: particle diffraction

Consider the single slit apparatus shown in Fig. 5.4. The apparatus consists of a particle source, a wall with a slit, and a detection screen. This is the same experimental configuration used earlier to demonstrate Newton's First Law. Here, however, the slit is much smaller and the particle is an entangled thing, possibly an electron. If we are to see any quantum effects, the slit width must be comparable in size to the electron's corresponding deBroglie wavelength.

Classically, we imagine that the electron is a free particle with no forces acting on it once it has left the source. As shown earlier, an ordinary particle always travels in a straight line and hits the detector at $\theta = 0°$. In classical physics, there is no such thing as particle diffraction. In the real world, diffraction only occurs with waves.

But, of course, this is not an ordinary experiment and we are not surprised that the electron can be deflected at some angle other than $\theta = 0°$. The electron is deflected without benefit of a deflecting force! And, more non-classical still, such an experiment is indeterminate. On another measurement attempt, there is a good chance that the electron strikes the screen at a different angle, even though we have not made any changes. Truth is, we have no idea where an electron will strike the screen. The quantum formalism makes no such predictions. Repeating the experiment many times the electrons are randomly scattered all over the detection screen at all possible scattering angles $-\pi/2 < \theta < \pi/2$. A possible angular distribution for such single slit diffraction is given in Figure 5.5. Note the maxima and minima that we associate with an interference pattern.

We do have a quantum explanation, of sorts, in terms of the Heisenberg Uncertainty Principle. For the single slit experiment considered here, the localizing slit introduces a position uncertainty Δy, which, according to the Heisenberg Uncertainty Principle, is accompanied by an uncertainty in momentum, $\Delta p_y \geq \dfrac{\hbar}{2\Delta y}$. Since $p_y = p\sin\theta$, the scattering angle θ is also indeterminate; the presence of the slit generates $\Delta\theta$, an uncertainty in the scattering angle. Consequently, there is the possibility of the particle being scattered at any angle. Although most electrons are scattered at small angles, some electrons are scattered at angles far from $\theta = 0°$.

A particle beam spreads out from the slit in all directions as a consequence of the uncertainty in momentum. Diffraction is not necessarily a wave property. In fact, diffraction experiments with particles are routinely done in present day labs.

SINGLE SLIT DIFFRACTION

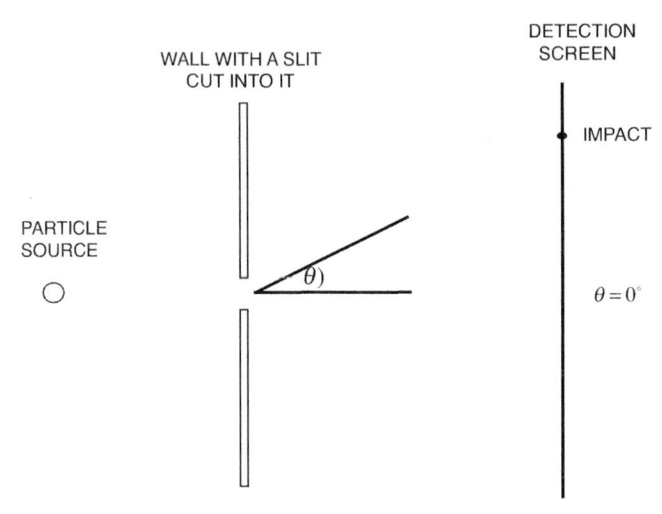

Figure 5.4. A single slit quantum experiment. A free particle is deflected without benefit of a deflecting force. Rather, it is the uncertainty in momentum, which accompanies the uncertainty in position introduced by the localizing slit, that gives rise to particle diffraction. Repeating the experiment generally yields a different scattering angle. We do not show a trajectory because there is none.

We see that the Heisenberg Uncertainty Principle is the quantum analog of Huygen's principle, which states that every point on a wavefront spreads out continuously in all directions. In a similar way, particle diffraction is explained in terms of the Heisenberg Uncertainty Principle: A particle localized at the point $y = y_0$ has an infinite uncertainty in momentum. Then, in a momentum measurement, there is equal probability of the particle being scattered in any direction. The particle beam spreads out discontinuously, one particle at a time.

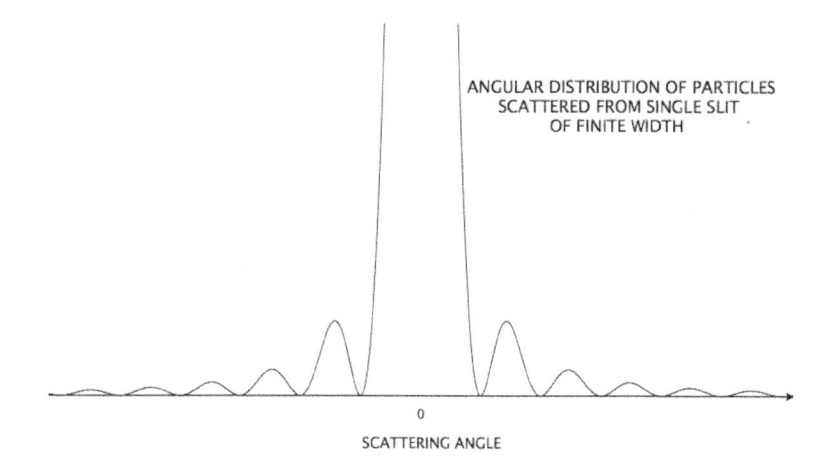

[Graph: ANGULAR DISTRIBUTION OF PARTICLES SCATTERED FROM SINGLE SLIT OF FINITE WIDTH, with horizontal axis labeled SCATTERING ANGLE centered at 0]

Figure 5.5. This is a plot of the probability function for the single slit experiment where the slit width is of the order of the deBroglie wavelength. Particles are scattered at angles other than zero. Note that the angular distribution of scattered particles shows evidence of interference maxima and minima. Particle diffraction is a consequence of the Heisenberg uncertainty principle.

6

QUANTUM INTERFERENCE
The role of entanglement in particle interference

We choose to examine a phenomenon which is impossible, absolutely impossible, to explain in any classical way, and which has in it the heart of quantum mechanics. In reality it contains the only mystery.

Richard P Feynman

6.1 General

In the final chapters of this book, we will describe experiments that allow us to compare the classical description with its quantum counterpart. In this chapter, we show that the classical approach fails to explain particle interference. In the real world of our everyday experience, particles are not supposed to do such things.

Yet, for nearly one hundred years now, we have seen experiments done with particles that do exhibit wave-like behavior.

The quantum interference experiments described here are done with particles. We profess that particle interference is possible when entanglement is in effect. All interference experiments require alternative paths through the apparatus. But in the quantum experiment, the alternative paths are entangled and have lost their individual identities. It is this loss of 'which way' paths that gives rise to particle interference.

6.2 Particle Interference

A classical description of interference in terms of waves was available to us long before the invention of quantum mechanics. If quantum particles were simply waves, then there would be no need for a quantum description of particle interference. We would need only define interference to mean 'wave interference' and be done with it. Some would have it that way.

In its everyday usage, the very word 'interference' implies two or more waves coming together in the same vibrating medium to generate an interference pattern. Destructive and constructive interference are created when the two waves are combined. In some places the waves are out of phase and cancel to give a resultant wave intensity of zero. In others the waves are in phase and the intensity is enhanced. We have seen this with water waves and heard it in sound waves. Classical particles cannot do this.

The equations of Newton and Maxwell do not support particle interference. There is no such thing in the real world. Real particles do not 'interfere' with each other. They might collide and scatter off one another, or maybe, stick together after the collision, but no one has ever seen two particles interfering in the manner of waves. It just doesn't happen. Feynman says it most forcefully, "It is impossible, absolutely impossible to explain" particle interference in any classical way.

Particle interference is a purely quantum phenomenon far removed from classical wave theory. The particles involved are obviously not ordinary things and any discussion of particle interference is outside the realm of classical physics. In what follows, we will show that particle interference is a consequence of quantum entanglement and the subsequent loss of the particle's trajectory, making it impossible to know which path was taken through the apparatus.

In the interference experiments to be described here, there are no waves propagating through the apparatus. There are only quantum particles, those entangled things that we have been discussing all along. Having come this far, you know such things do not obey the classical laws.

Diffraction and interference experiments done with particles are now rather common. It is an empirical fact that some experiments done with particles do have the characteristics associated with classical waves. The single slit diffraction experiment discussed earlier is such an example.

In fact, interference experiments have been done with C_{60} and even with heavier molecules. There is little doubt that these relatively massive objects satisfy our definition of 'particle'. Yet, under certain conditions, they too exhibit diffraction and interference.

Particle interference requires two conditions:

1. The apparatus must be set up in such a way that the particle has more than one route through the apparatus to the detector.
2. Entanglement must be in effect, making it impossible, absolutely impossible, to determine which route is taken. The particle has no trajectory.

If these two conditions are met, then the experiment will exhibit particle interference. The classical maxim, "Waves exhibit interference. Particles do not." no longer applies. Rather, we now say, "Some experiments done with particles exhibit interference. Others do not." It is the entangled experiment, not the particle itself, that has wave-like properties.

The experimental configuration must provide alternate routes through the apparatus. These alternate routes are real paths in real space. We can see them in the laboratory. Each route is a possible trajectory, a real characteristic of a moving classical particle. As long as we can distinguish between the alternate routes, classical physics reigns and there is no interference.

But entangled particles do not exist in space-time where trajectories exist. Consequently, there are no paths through an entangled apparatus. The alternate routes through the apparatus are now lost in the entangled whole.

It is important to understand that the particle must not have a trajectory if interference is to occur. We do not mean that the particle does have a trajectory, but, we just don't happen to know where it is. This is not merely human ignorance at play here. The particle truly does not have a trajectory! In the quantum experiment there is no particle, separate from everything else, moving through the entangled apparatus on its way to the detector.

Entanglement excludes trajectories and makes interference possible.

6.3   A generic interference experiment with particles

We now construct a generic particle interference apparatus in which the two conditions for interference are satisfied. This experiment consists of a particle source, a beam splitter, an inverse beam splitter, and two detectors. We will count the number of particles that reach each detector. With entanglement in effect, we expect to see particle interference.

As usual, our particle source emits particles so slowly that, at most, only one particle is ever present in the apparatus at any time. Present day particle sources have this capability. Since there is never more than one particle anywhere in the apparatus, there is no possibility of any recombining, or interfering, of particles.

The beam splitter shown in Figure 6.1 does just what its name implies. It splits a beam into two beams. It has one input and two output channels. It provides two alternate paths for a classical particle to get through the experimental apparatus to a detector. Which route an individual particle takes is a purely random event. There is no reason why one particle goes one way and another goes the other way and it is impossible for us to predict which way a given particle will go. As we have seen, a Stern-Gerlach magnet with spin 1/2 particles is such a device.

We assume that this beam splitter does not favor one route over the other. One half of the incident particles are in Route A and the other half of the incident particles ends up in Route B. We can confirm this by counting the number of emergent particles in each detector. All particles that reach Detector 1 are from Route A and all particles in Detector 2 have traveled via Route B.

We find the probabilities:

$P_A = \dfrac{1}{2}$ is the probability that an incident particle travels Route A.

$P_B = \dfrac{1}{2}$ is the probability that an incident particle travels Route B.

BEAM SPLITTER

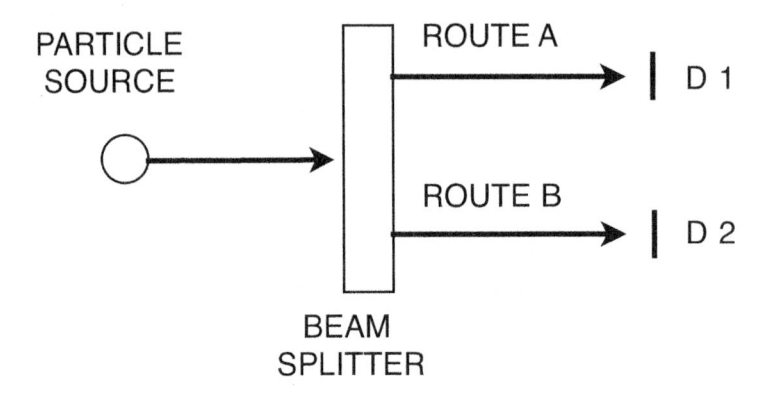

Figure 6.1. A beam splitter is a quantum device that splits the particle beam in two. Half of the incident particles take Route A into Detector 1 (D1), and the other half ends up in Detector 2 (D2) by way of Route B. There is no reason why a particle goes one way or the other. Where the particle ends up is a purely random event.

All this is undone with the inverse beam splitter in place. It is the the inverse beam splitter that 'entangles' the entire apparatus into a whole. It is often identical to the beam splitter, but now, there is no way to determine which way a particle actually went on its way through the apparatus. Entanglement is in effect and the particle does not have a trajectory. In particular, Routes A and B become entangled and their individual identities are lost. There is no obvious way to get through the apparatus. Even Mother Nature doesn't know which way the particle went. The second condition is satisfied and interference is possible. This generic interference experiment is shown in Figure 6.2.

INTERFERENCE EXPERIMENT

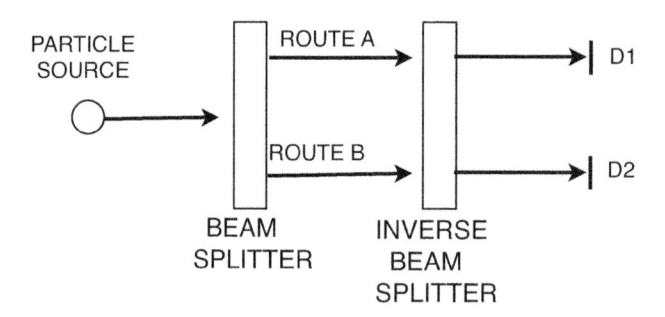

Figure 6.2 This is the generic interference experiment consisting of a particle source, a beam splitter, an inverse beam splitter, and two detectors. With the inverse beam splitter in place all parts are entangled into a single entity. There is no longer a Route A and no longer a Route B, in spite of what we see in the diagram.

A particle emerging from the inverse splitter is no longer associated with a given path. Entanglement makes it impossible to answer the question, "Which way did the particle go?".

Classically, with both routes open, each route contributes to the total number of particles in each detector. Some of the particles in Detector 1 come from Route A and others would come from Route B. Our intuition tells us that in order for a particle to get through the apparatus, it takes either Route A or it takes Route B. There is no other choice. Classically, it has to go one way or the other. Just common sense, all this.

To verify this classical assumption, let us analyze each route separately. With the inverse beam splitter in place, we block Route B.

Now, all particles reaching the detectors are necessarily from Route A, since it is the only open route. We know which way the particle went. We do not expect any interference.

This is a real experiment with separability in effect. All parts of the experiment are real including the particle. We call this Real Experiment A. It is shown in Figure 6.3.

In Real Experiment A, we find that one half of the Route A particles reach Detector 1 and one half reach Detector 2. Since only one half of the incident particles take Route A to begin with, we conclude that the probability that an incident particle takes Route A on its way to Detector 1 is 1/4. The probability that an incident particle takes Route A on its way to Detector 2 is also 1/4.

REAL EXPERIMENT A

$P_A(1) = \frac{1}{4}$

$P_A(2) = \frac{1}{4}$

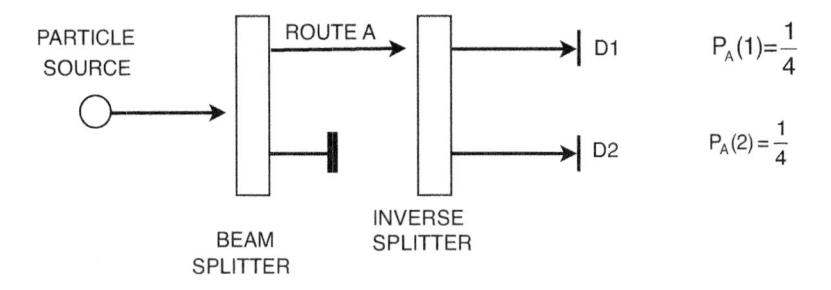

Figure 6.3. Route B is blocked. We are certain that all particles that reach the detectors must have come via Route A, since it is the only possible way for a particle to traverse the apparatus. We find that 1/4 of the incident particles end up in Detector 1 (D1) and 1/4 of the incident particles reach Detector 2 (D2).

We write the results obtained in Real Experiment A:

$P_A(1) = \dfrac{1}{4}$ is the probability that an incident particle takes Route A to Detector 1.

$P_A(2) = \dfrac{1}{4}$ is the probability that an incident particle takes Route A to Detector 2.

We then block Route A, as shown in Figure 6.4, and repeat the same procedure. Here, all detected particles must have traveled Route B. We call this Real Experiment B.

REAL EXPERIMENT B

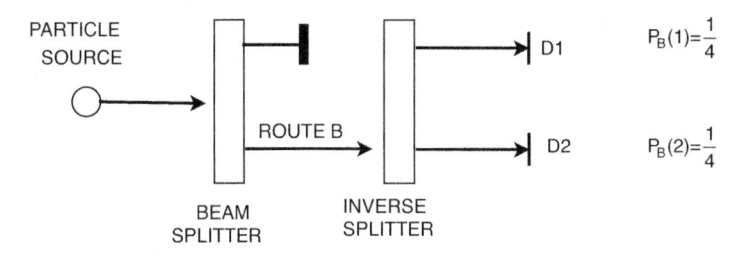

Figure 6.4. Route A is blocked. We are certain that all particles that reach the detectors must have come via Route B. Now, Route B is the only way through the apparatus. We find that 1/4 of the incident particles end up in Detector 1 (D1) and 1/4 of the incident particles reach Detector 2 (D2).

We obtain the same probabilities with Route A blocked:

$P_B(1) = \dfrac{1}{4}$ is the probability that an incident particle reaches Detector 1 via Route B.

$P_B(2) = \dfrac{1}{4}$ is the probability that an incident particle reaches Detector 2 via Route B.

Note that the total probability is
$$P_A(1) + P_A(2) + P_B(1) + P_B(2) = 1.$$

We have accounted for all the incident particles. As suspected, a particle takes either Route A or Route B on its way to the detectors. Nothing out of the ordinary here! This is classical physics applied to each of two mutually exclusive experiments.

We now use the results of these two real experiments to predict the results of Experiment AB, where both routes are open, as shown in Figure 6.5. The classical physicist is confident that Experiment AB is also real and separable. Classically, both routes open means that there is a real Route A and a real Route B through the apparatus. Each route is available to the particle.

The realist truly believes that some particles take Route A and others take Route B on their way to the detectors. Why shouldn't they? Both routes are still open and a particle will naturally go one way or the other. Each route independently provides a way for the particle to get through the apparatus. Simply, common sense!

The realist knows how Experiment AB works. It functions according to the rules of classical physics. Each individual route contributes particles to each detector.

Further, he believes that the principle of locality is in effect. Whenever a particle takes Route A, it doesn't matter if route B is open or closed.

The results obtained in Experiment A should still be valid with both routes open, even though Experiment A and Experiment AB are mutually exclusive. After all, particles moving through Route A are free and independent of what is happening to Route B. Opening or closing Route B has no effect on the behavior of particles in Route A. Locality does not allow any changes made to Route B to have any effect on particles in Route A. Voodoo physics does not work here! The only effect we expect when we open Route B is to see more particles getting through the apparatus and those new Route B particles will add to those from Route A. If we close Route B, fewer particles can get through the apparatus.

EXPERIMENT AB (INTERFERENCE)

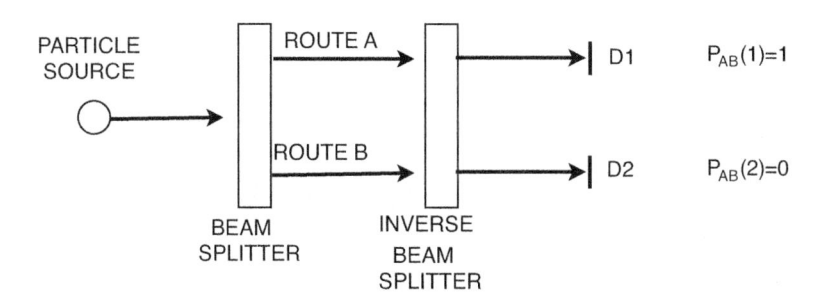

$P_{AB}(1)=1$

$P_{AB}(2)=0$

Figure 6.5. Interference with particles. The apparatus is now an entangled whole, in spite of what we see in this diagram. With both routes open, all particles end up in Detector 1 and no particle ever reaches Detector 2. This is constructive interference in Detector 1 and destructive interference in Detector 2.

It seems obvious that Routes A and B are still two possible ways for particles to get through the apparatus when both routes are open. That being the case, Experiment AB is also a real experiment and there is no interference.

We can confirm this by counting the number of particles in each route separately, as done in Real Experiments A and B, or simultaneously, as in Experiment AB. It doesn't matter! In classical physics, the probabilities add. Thus, the probability that a particle reaches a detector in Experiment AB is

$$P_{AB} = P_A + P_B,$$

where P_A is the probability, obtained in Real Experiment A, that a particle reaches a detector via Route A. Likewise, Real Experiment B gave us the probability P_B.

We can now predict how many particles reach each detector in a real Experiment AB. Adding the probabilities obtained in the earlier experiments, classical physics predicts that 50% of all incident particles reach Detector 1. Likewise, the same classical reasoning predicts that 50% of all incident particles are found in Detector 2.

Unfortunately, this is all wrong! Everything we have said up to this point is how we imagine a classical experiment would function.

It turns out that in Experiment AB, with both routes open, all particles are in Detector 1. No particle ever reaches Detector 2. We have constructive interference in Detector 1 and destructive interference in Detector 2. The apparatus of Experiment AB is actually an entangled interference experiment and the rules of classical physics no longer apply. There is no classical explanation.

Our classical predictions are correct for Real Experiments A and B individually, but are obviously in error for the combined Experiment AB. Obviously, we did not tell you that the experimentalist had set up her apparatus so that constructive and destructive interference would be evident.

So, where did we go wrong? With both routes open and the inverse beam analyzer in place, we assumed that there are still two possible routes for a particle to follow. Our intuition has led us astray. Actually, with entanglement, there are no paths through Experiment AB. If there were, then there would be trajectories along with no interference.

With the inverse beam splitter in place, the individual routes, along with all other parts of the entangled apparatus, lose their individual identities. Even though both routes seem to be open, they are no longer accessible to the particle. How can that be? We now know that 'routes' and 'particles' are classical concepts that imply 'separate particles' and 'separate routes'. But they are not separable in the entangled whole. We should stop talking about them! But, of course, that is impossible for us to do.

It is the entangled experiment that makes particle interference possible.

So, how does the particle get from the source through the apparatus to a detector? We have no idea! There is no answer to such a question. Remember Wheeler. There is no space-time description of an individual particle working its way through the apparatus. Quantum mechanics denies any such thing. Rather, we should be talking about the entangled whole, which has no inner workings. We must continually remind ourselves that there are no separate particles moving through an entangled apparatus. If we assume otherwise, as we just did, then our predictions are wrong.

Repeating Experiment AB many times, 100% of all incident particles are in Detector 1. No particles are ever found in Detector 2. That is all there is to know about it! Quantum mechanics does not provide any further details and there is no need for us to know anything more.

It does not matter how many alternate routes there are, or how many detectors we have. We have used two detectors because we want this experiment to be the model for other experiments to be discussed shortly. All interference experiments have the same characteristics as this basic two-detector set-up.

But, what happened to all those particles in Real Experiments A and B that took Routes A and B on their way to Detector 2? Although this seems like a reasonable question, it is not. It is meaningless because those earlier results were from mutually exclusive experiments done with other particles at other times. The probabilities P_A and P_B have nothing to do with Experiment AB. We did not obtain those probabilities from Experiment AB and, therefore, they are not applicable to Experiment AB. When we do Experiment AB, Real Experiment A and Real Experiment B are unperformed experiments that have no results.

There is no such thing as P_A, and there is no P_B when we do Experiment AB. Just because some particles happen to reach Detector 2 in some other experiments there is no reason to expect that some particles in Experiment AB will do the same. That is counterfactual reasoning and it is wrong. We must discipline ourselves to reject such classical thinking. Admittedly, not an easy task.

The figures shown for Real Experiments A and B are correct, but Fig. 6.5 for the entangled Experiment AB is not. Entangled Experiment AB does not have a pictorial representation. Figure 6.5 only shows how we assembled the apparatus from real devices.

The entanglement of the routes removes any possibility of a particle having a trajectory and both requirements for interference are evident. We are now in the quantum domain where particle interference is possible.

It is time for us to give a quantitative approach to particle interference. We will show that particle interference is inherent in the quantum formalism itself. Once again, a bit of math is required.

6.4   The superposition of probability amplitudes

We now know that a quantum experiment yields only the possible measurement results a_k, say, and the probabilities $P_k = |\psi_k|^2$ for obtaining each result. ψ_k is the probability amplitude corresponding to the probability P_k. In general, the probability amplitude is a complex number.

Consider, again, the three experiments discussed above. We expected the probability that a particle reaches a detector in Experiment AB to be $P_{AB} = P_A + P_B$. But, that didn't work! Adding probabilities is how we do classical physics. We now calculate the probability the correct quantum way.

In quantum mechanics we do not add the probabilities. Rather, we add the complex amplitudes. The general rule for calculating the probability is:

Always add the amplitudes for the alternate routes considered separately before squaring them to get the probabilities:

$\psi_{AB} = \psi_A + \psi_B$, and $P_{AB} = |\psi_{AB}|^2 = |\psi_A + \psi_B|^2$.

This general rule for calculating the probability is called "the superposition of amplitudes".

The probability amplitudes ψ_A and ψ_B are complex numbers from which we calculate the corresponding probabilities $P_A = |\psi_A|^2 = \frac{1}{4}$ and $P_B = |\psi_B|^2 = \frac{1}{4}$.

Any complex number can be written in terms of an imaginary exponent. Thus, the complex amplitude for a particle to reach a detector in Real Experiment A has the general form

$$\psi_A = |\psi_A| e^{i\phi_A} = \frac{1}{2} e^{i\phi_A}.$$

Here, $e^{i\phi_A}$ is called a phase factor and ϕ_A is the phase angle. The phase factor has no effect on the probability $P_A = |\psi_A|^2 = \frac{1}{4}$. Likewise, the complex probability amplitude in Real Experiment B is

$$\psi_B = |\psi_B| e^{i\phi_B} = \frac{1}{2} e^{i\phi_B}$$

and its corresponding probability is $P_B = |\psi_B|^2 = \frac{1}{4}$.

Adding the amplitudes as directed in the general rule gives

$$\psi_{AB} = \psi_A + \psi_B = \frac{1}{2} \left[e^{i\phi_A} + e^{i\phi_B} \right]$$

and the corresponding probability is

$$P_{AB} = |\psi_{AB}|^2 = |\psi_A + \psi_B|^2 = \frac{1}{4} \left[e^{i\phi_A} + e^{i\phi_B} \right] \left[e^{-i\phi_A} + e^{-i\phi_B} \right].$$

A bit of algebra gives

$$P_{AB}(\phi_B - \phi_A) = \frac{1}{2} \left[1 + \cos(\phi_B - \phi_A) \right].$$

Note that any interference effects depend on the relative phase $\phi_B - \phi_A$, and not on the individual phase angles.

Entanglement makes particle interference possible while the relative phase difference determines where constructive and destructive interference occurs.

6.5 Constructive and destructive interference

When $\cos(\phi_B - \phi_A) = 1$, the corresponding probability is

$$P_{AB}(\phi_B - \phi_A) = 1.$$

All the incident particles are in that detector. We have constructive interference.

Likewise, when $\cos(\phi_B - \phi_A) = -1$, then

$$P_{AB}(\phi_B - \phi_A) = 0.$$

No particle ever reaches that detector. This is destructive interference.

In the entangled Experiment AB, the apparatus was set up such that $P_{AB}(\phi_B(1) - \phi_A(1)) = 1$ and $P_{AB}(\phi_B(2) - \phi_A(2)) = 0$.

All the incident particles are found in Detector 1.

There is absolutely no way to explain what is going on here. Just as it is with other quantum experiments, we do not know how the interference experiment works. Bohr says there is nothing to know, so we should not concern ourselves with it any further.

All we can say is that in entangled Experiment AB all particles are in Detector 1 and no particle is ever in Detector 2. There is no realistic explanation in terms of the classical laws.

Particle interference is simply another consequence of quantum entanglement.

6.6 Complementarity and quantum interference

A trajectory is incompatible with interference. They are complementary concepts. We can do classical experiments where particles have trajectories, or we can do quantum experiments that exhibit particle interference, but we never see both in the same experiment. It is a well known empirical fact that 'which way' information stifles interference.

If somehow it is possible, in principle, to determine which path the particle took, then the particle has a trajectory and there is no interference. This does not mean that the human observer must know which way the particle went. Usually, he doesn't know. Nevertheless, there is no interference.

We can do interference experiments with photons, electrons, and neutrons provided there is no trajectory to reveal which way the particle went. As strange as it seems, we have no idea how the particle gets to the detector in an interference experiment.

If we did, there would be no such thing as particle interference. Although the interference apparatus provides alternative paths, no path is accessible to the particle. Entanglement makes it impossible. If there is to be interference, there can be no identifiable way for a particle to get through the apparatus to the detector!

This can be perplexing because we imagine a particle moving through the experimental apparatus on its way to the detectors. It seems obvious, with or without interference, that the particle requires a pathway through the apparatus. At least in classical physics it does.

Numerous experiments have shown that interference does not occur in any experiment that reveals which way the particle went. In Real Experiments A and B we have identified the route taken by blocking the route not taken. But in entangled Experiment AB it is impossible to identify which way the particle went

All three experiments discussed above are mutually exclusive and their results are complementary. We cannot block one of the routes and still have both routes open at the same time. It is physically impossible.

The classical attributes of the two real experiments do not carry over into the entangled interference experiment. Remember, we must not combine the results of mutually exclusive experiments. Knowing that it is possible for a particle to take Route A in Real Experiment A, does not mean another particle can still take Route A in Experiment AB. That is counterfactual thinking and it is wrong! The classical particle behavior observed in each of the two real experiments is no longer present in the interference experiment.

Let's assume that, with both routes open, the particle does go one way or the other, as classical particles are wont to do, but, it just so happens that we don't know which way. It doesn't matter whether we know or not. If a particle does go via Route A, for example, then it has a trajectory and there is no interference.

This has been demonstrated in many experiments where ingenious experimenters know, in principle, which way the particle went, even with all paths open. In all such cases, there is no interference.

In the interference experiment, then, the particle does not go by way of Route A. It does not go by way of Route B, either.

So which way does it go? There must be a way for it to get to the detector, isn't there? Don't these 'particles', whatever they are, have to travel somehow through the apparatus on the way from source to detector? No, they do not! That is how classical particles do it. But, again, we emphasize that these are not classical particles moving through the apparatus and we should not expect such behavior.

But, maybe the particle takes both paths simultaneously? Classical waves do. But these are not classical waves. If we do an experiment in which we know that a particle takes Route A, then we know for sure that the particle is not in Route B. The orthogonality of the eigenvectors prevents it. A particle cannot go both ways simultaneously.

We are forced to conclude that there is particle interference only if we reject the notion that there is a way for the particle to get through the apparatus.

Routes A and B are lost in the entangled whole. Obviously, the space-time description of classical physics does not apply to the entangled interference experiment.

If quantum particles did have trajectories, there would be no such thing as particle interference. There would be only wave interference and we would not be having this discussion.

Remember Wheeler's admonition. We do not have a space-time description for quantum particles. We should not be talking about a particle taking this route or that through the apparatus. We need to stay focused on the entire entangled arrangement at the instant that the particle is detected, not on an individual particle.

The problem is that we do talk about an entangled particle as if it was an ordinary thing. It is not ordinary. The unpleasant truth is, in the entangled whole there is no ordinary particle, separate from everything else, moving through the apparatus. Otherwise, there would be no particle interference.

6.7 The Mach-Zehnder interferometer

Here, we demonstrate particle interference with a Mach-Zehnder interferometer, which is identical to Experiment AB. This is an actual experiment done with photons. It can be found in many undergraduate physics labs.

Again, we assume the light source is so dim that only one photon, at most, is in the apparatus at any instant. We will not measure energy or momentum, so there is never any reference to frequency or wavelength. This is not an experiment done with light waves. It is all about an experiment done one photon at a time.

The apparatus consists of a photon source, two beam splitters, two mirrors and two detectors configured as shown in Figure 6.6. Here, the beam splitter is a partially reflecting mirror that transmits half of the incident photons and reflects the other half. It splits a beam of photons, not individual photons. There is no such thing as a half photon. All the photons in the detectors are identical whole photons. Photons reflected from a beam splitter look just like transmitted ones.

We imagine that the first beam splitter provides the two alternate routes for a photon to follow on its way through the apparatus.

MACH-ZEHNDER EXPERIMENT

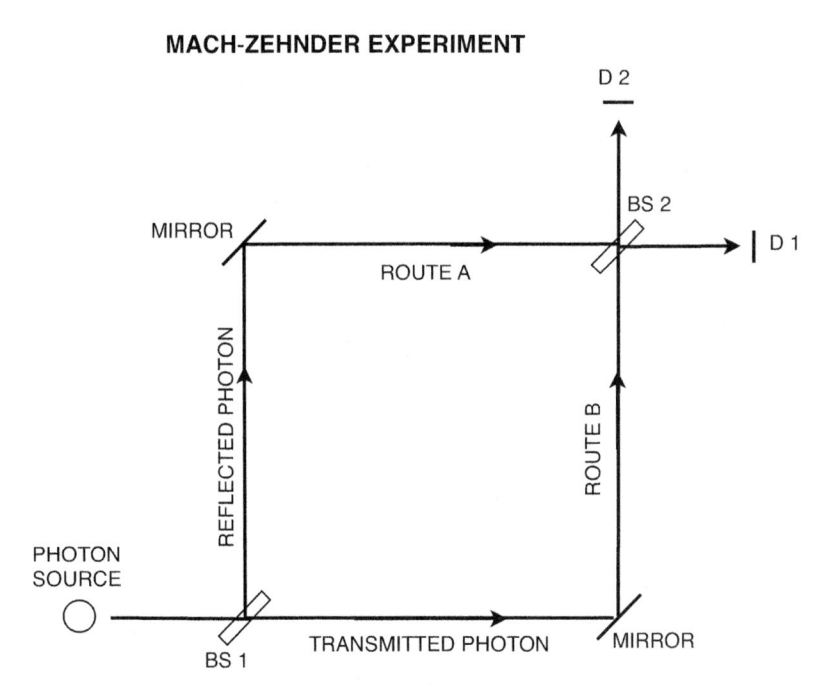

Figure 6.6 The Mach-Zehnder interference experiment is identical to Experiment AB. With the second beam splitter (BS2) in place there is interference even though there isn't anything for a single photon to interfere with. With BS 2 removed, all the Route A photons are in D 1 and all Route B photons are in D 2. When the apparatus is set up with BS 2 in place, all the incident photons end up in Detector 1 (D 1). Strange as it seems, no particles ever reach Detector 2 (D 2).

The second beam splitter is the inverse beam splitter for this apparatus. With the second beam splitter in place, the two routes are lost in the entangled whole. We do the Mach-Zehnder experiment with a single photon, so there is nothing for a photon to 'interfere' with at the second beam splitter. Nevertheless, because the second beam splitter randomly transmits or reflects the individual photon, entanglement is in effect and particle interference is possible.

Everything done in the three experiments above can be performed with the Mach-Zehnder apparatus. We can block each route separately or have both routes open.

Assuming each beam splitter reflects particles and transmits particles equally, we again have $P_A = P_B = \frac{1}{4}$ at each detector. With both routes open, common sense (read 'classical physics') tells us that a photon must take one route or the other and the separate probabilities for reaching Detector 1 are added: $P_{AB}(1) = P_A(1) + P_B(1) = \frac{1}{2}$. This is what a classical physicist would do. Likewise, classical reasoning tells us the probability for reaching Detector 2 with both routes open is $P_{AB}(2) = P_A(2) + P_B(2) = \frac{1}{2}$. We expect no interference.

But, this is not what happens. Classical physics does not work here! With both routes open, the apparatus was set up so that all of the photons are in Detector 1! None of the photons ever reach Detector 2. We get $P_{AB}(1) = 1$ and $P_{AB}(2) = 0$. We see constructive interference at Detector 1 and destructive interference in Detector 2.

Again, counterfactual reasoning, so effective in classical thought, gets us into trouble. This is not a classical experiment and we should not be discussing it as if it were. Even though we call them 'particles', photons are not ordinary and they do not obey the particle physics of Newton and Maxwell.

We can correctly predict the experimental results only if the routes are not separate entities. We can explain the interference with particles only if there is no longer any photon transmitted from the first beam splitter. If there was, then we would know which way the photon went, and there would be no interference. A transmitted photon is incompatible with interference because it defines a trajectory. Likewise, there are no reflected photons from the first beamsplitter. There are no separate entities in the entangled whole. Figure 6.6 shows a classical apparatus consisting of separate devices. The quantum experiment has no such pictorial depiction.

The problem, as discussed above, is that we do talk about this experiment as if the photon traverses the apparatus from source to detector, undergoing a series of encounters with beamsplitters, mirrors, and detectors. This is true for ordinary particles moving through space-time in a classical apparatus. And, the common language we use misleads us into thinking that photons also behave the same way. They don't!

All we know for sure is that when we do this experiment there is always a photon in Detector 1. No photons are ever found in Detector 2. That is the nature of this particular set up. The postulates allow us to determine the probabilities of reaching each detector and that is all that should concern us. Any discussion of a photon going this way or that is meaningless. It only adds to the confusion. But, unfortunately, we do it all the time. We incorrectly speak as if separability and locality are still in effect.

But, many ask, how are we supposed to describe photon behavior if we are not allowed to talk about photons? The point is that we do not discuss photon behavior because that is not what quantum mechanics is about. Rather, we are describing an interference experiment done with photons. This is not just a play on words. We must learn to talk about the experimental whole, which is actually what we are describing.

If we want to talk of photons moving through the apparatus, then we must actually go and do a classical experiment where 'photons are moving through the apparatus'.

As discussed earlier, we cannot escape from our classical mindset. We still use classical concepts and ordinary language when discussing these things, even though they no longer have any meaning. The description given here of the Mach-Zehnder experiment implies that it is a classical apparatus consisting of separate devices that operate in ordinary ways. We must reject such thinking and remind ourselves, again and again, that an entangled photon is a quantum thing that exists only in conjunction with the entire apparatus.

An entangled photon and an entangled whole are one and the same in the quantum description. Questions, such as, "Which way did the photon go? Did it go both ways?" are meaningless. There are no answers to such questions. Quantum mechanics does not describe such things.

What an individual photon is doing in the entangled apparatus before reaching the detector is not part of the quantum discussion. If you seek answers to such questions then you must do classical experiments.

All we know is that this particular Mach-Zehnder apparatus, with both routes open, does exhibit particle interference. We observe constructive interference in Detector 1 and destructive interference in Detector 2. That is all that we should be saying about it!

6.8 The double slit experiment

No discussion of quantum interference would be complete without the double slit experiment done with particles.

The double slit experiment, shown in Figure 6.7, consists of a particle source, a wall with two slits, and a detection screen. The slits correspond to the first beam splitter in the above experiments. We assume that the slits are infinitesimally thin, so that the slits provide only two possible routes through the wall. Classical particles would pass through one of the slits on their way to the detection screen.

DOUBLE SLIT INTERFERENCE EXPERIMENT

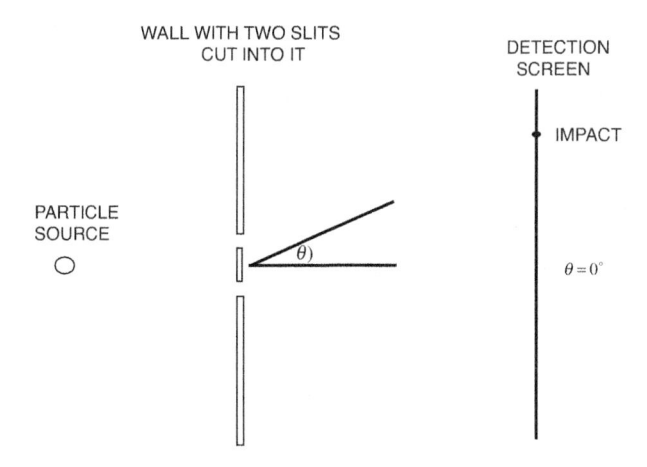

Figure 6.7. This is Young's experiment done with free particles. The two slits correspond to the beam splitter in the generic interference experiment. Particles impact the detection screen one at at a time and it is the statistical distribution of many scattered particles that is identified as an interference pattern.

In the earlier interference experiments discussed here, there were only two end detectors and only two possible values of the phase difference $\phi_B - \phi_A$, one for each detector. But, slits are localizing devices that introduce an uncertainty in the particle's position.

According to the Heisenberg Uncertainty Principle there is a corresponding uncertainty in momentum and, consequently, particles are scattered from the slits in all possible directions. It is the Heisenberg Uncertainty Principle that makes it impossible to determine which slit the particle went through. Thus, it is the slits themselves along with momentum uncertainty that enables entanglement.

In particle interference we do not see bright and dark interference fringes until many particles have been accumulated on the detection screen.

Although the particles are scattered randomly, some scattering angles occur more frequently while some other directions are excluded entirely. The preferred directions are where constructive interference occurs. No particles are ever seen in those directions where destructive interference occurs.

What we call an interference pattern, then, is the statistical distribution of many different particles hitting the screen at different times. The individual particles impacting the screen are always localized as dots and we cannot discern any wavelength or frequency from a single dot.

If this were wave interference we would always see a continuous smear across the detection screen, even when the light is extremely dim. We would never see any dots. Rather, the intensity of the entire interference pattern would be gradually lessened as the light is dimmed.

As in previous experiments, we assume an equal number of particles from each slit reach the detection screen one at a time. But, here, a scattered particle can hit at any point on the screen. It is as if there are an infinite number of detectors and there is a corresponding infinite number of values for the relative phase angle $\phi_B - \phi_A$, one for each possible impact point. The relative phase angle $\phi_B - \phi_A$, then, is a continuous variable and the probability becomes a periodic function of $\phi_B - \phi_A$.

As usual, constructive interference occurs on the screen where $\cos(\phi_B - \phi_A) = 1$ and the corresponding probability function is $P_{AB}(\phi_B - \phi_A) = 1$, its maximum value. As always, destructive interference occurs where $\cos(\phi_B - \phi_A) = -1$, and the probability function is $P_{AB}(\phi_B - \phi_A) = 0$.

A plot of the probability function in terms of the scattering angle is shown in Fig. 6.8.

ANGULAR DISTRIBUTION OF PARTICLES
SCATTERED FROM DOUBLE SLIT

SCATTERING ANGLE (RADIANS)

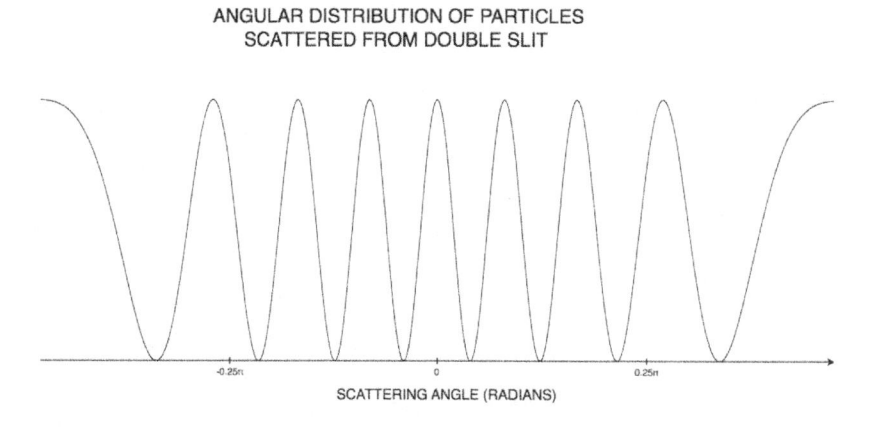

Figure 6.8. The angular distribution of particles scattered from double narrow slits. This is a plot of the probability function $P_{AB}(\phi_B - \phi_A)$ in terms of the scattering angle. The maxima and minima correspond to constructive and destructive interference, respectively.

Young's experiment, once considered to be definitive proof that light is a wave, now has a quantum explanation in terms of photons, as well as other particles. Light as photons does not preclude interference.

Some of you might argue that these last two experiments have perfectly good descriptions in terms of light waves that do predict the correct results. In such classical descriptions, the interference is wave interference, so the Mach-Zehnder interferometer and the double slit experiment have no need for the photon description given here. The classical wave description works just fine.

To answer such critics we now present a purely quantum interference experiment done with particles with no competing classical wave description.

6.9 Interference with a modified Stern-Gerlach experiment

Recognizing that a Stern-Gerlach magnet is a beam splitter, we now show interference effects obtained with spin 1/2 particles. This requires a modified Stern-Gerlach apparatus that consists of several Stern-Gerlach magnets oriented in the z-direction such that the two beams emerging from the first magnet are then recombined by these other magnets before leaving the apparatus.

With this modified apparatus, as shown in Figure 6.9, all Route A particles have $S_z = +1$ and all particles in Route B have $S_z = -1$. Here, we measure the z-component. Although there are two alternate routes for the particle to take, and although the beams are then recombined, we do not see any interference effects when the z-component is measured. This is because knowing the spin value identifies the route taken. When we get spin up we know the particle came via Route A. Likewise, spin down means the particle travelled Route B. We do know, in principle, which way the particle went.

As long as we measure the z-component of the spin vector, we know the route taken and there is no interference.

MODIFIED STERN-GERLACH EXPERIMENT

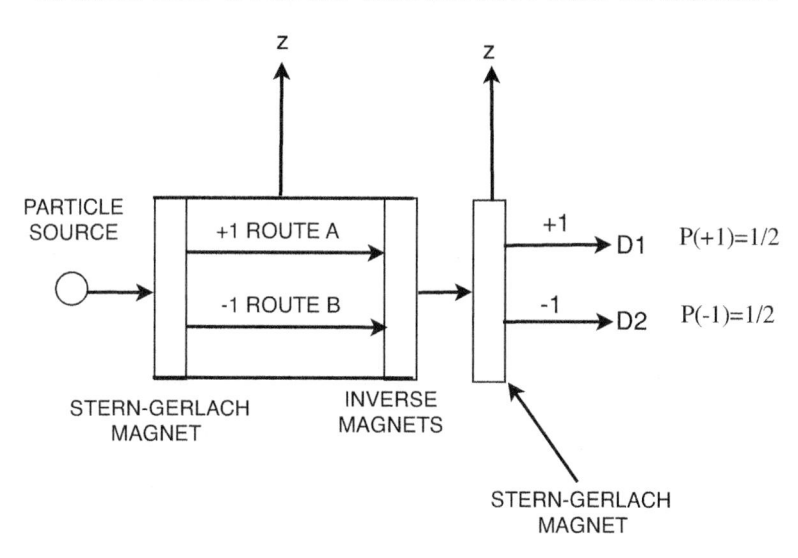

Figure 6.9. The modified Stern-Gerlach apparatus is equivalent to the Stern-Gerlach Experiment discussed in Section 4.7. Here, we measure the z-component of the spin 1/2 vector. All particles in D1 arrived via Route A with $S_z = +1$. Likewise, all particles in D2 came by way of Route B with $S_z = -1$. There is no interference because the spin value identifies which way the particle went.

But, we do observe interference effects by measuring the x-component of the spin instead of the z-component. We do this by inserting a Stern-Gerlach magnet oriented in the x-direction in front of the detectors, as shown in Figure 6.10.

All particles entering Detector 1 now have spin up in the x-direction and all particles in Detector 2 have spin down. These are spin values S_x, not to be confused with S_z that are associated with Routes A and B.

The x-and z-components are complementary variables, so measuring the spin in the x-direction makes it impossible to know the value of S_z. The Stern-Gerlach magnet in the x-direction plays the role of the required inverse beam splitter.

STERN-GERLACH VERSION OF INTERFERENCE EXPERIMENT AB

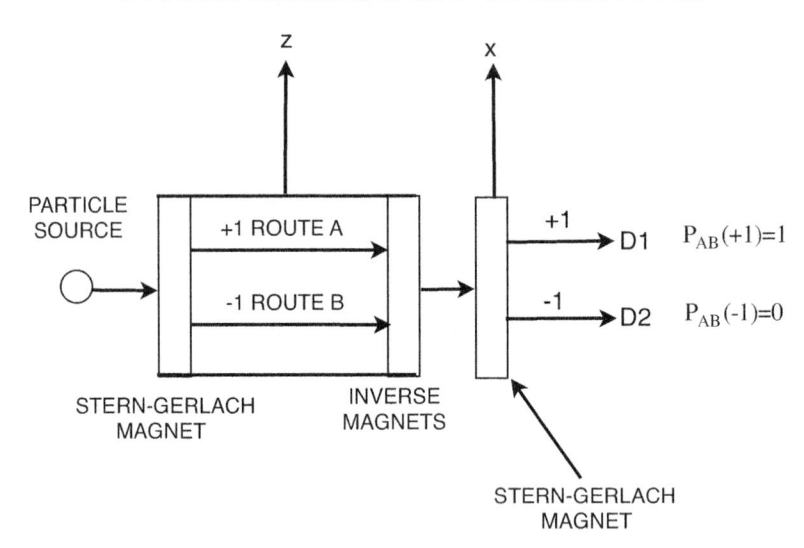

Figure 6.10. This is an interference experiment done with spin 1/2 particles. With both routes open, the entire apparatus is entangled into an experimental whole. Measuring the x-component does not reveal which way the particle went. There is no trajectory. We have constructive interference in Detector 1 where all particles have $S_x = +1$. No particles ever reach Detector 2.

When we measure the x-component, the entire apparatus is an entangled whole. Routes A and B through the z-apparatus are lost and we have no way of knowing which route was taken. Both conditions required for interference are satisfied.

We now have the required interference apparatus. The experimental set-up is now identical to the generic interference experiment discussed above. When we measure the x-component of particles emerging from the modified Stern-Gerlach apparatus in the z-direction, we expect interference.

As usual, we can make the classical prediction $P_{AB} = P_A + P_B$ from the probabilities obtained in the Stern-Gerlach versions of Real Experiments A and B, which are obtained by blocking the appropriate route. The realist, again, assumes that P_A and P_B are still applicable when we do the Stern-Gerlach version of the interference Experiment AB. He expects, then, that each route is still contributing half of the total number of particles reaching each detector. But, again, classical thinking is wrong.

When we actually do the experiment, all the detected particles have spin up in the x-direction! There are no particles with spin down. We have constructive interference in the spin up direction and destructive interference with spin down. We cannot replicate these results by adding the probabilities as mandated by the rules of classical physics.

Again, quantum interference is seen to be a consequence of entanglement and the subsequent loss of which way paths through the apparatus.

7

EINSTEIN, PODOLSKY, AND ROSEN
An attempt to show that quantum mechanics is incomplete

No reasonable definition of reality could be expected to permit this.

Einstein, Podolsky, and Rosen

Quantum mechanics is not a theory about reality.

Asher Peres

7.1 General

Einstein was one of the revolutionary developers of quantum theory. Nevertheless, he was the quintessential classical physicist and he remained so his entire life.

For Einstein, scientists are supposed to describe the universe as it is, an objective reality consisting of separable real things moving about in space-time. Classical experiments are constructed from individual parts and physicists are able to describe the behavior of each part separately, if so desired. Each part works as dictated by the classical laws, even when no one is paying any attention.

Classical experiments exhibit the principles of separability and locality. Separability means that each individual part has a specific location in space, spatially separated from other real things. A separate particle undergoes a sequence of events as it interacts with other parts of the apparatus. We know what is going on inside the classical experiment before the particle reaches the detector. We can explain how the result comes about and there is a reason why we get one result instead of another.

Locality means that all interactions are necessarily induced by contact forces. A particle can influence only particles at its immediate location without any mediating connection between them. 'Jumps' from one location to another are not allowed in the space-time continuum.

We are able to construct machines, instruments, vehicles and other things where all the individual parts are connected to make up a finished product. Most importantly, in principle at least, the finished product functions as it was designed to. In our everyday world, a composite system is the sum of its parts. We know how things work!

This classical approach is enormously successful, and Einstein was not about to give up on it. He saw no reason why atoms should behave any differently. He asserted that atoms, like everything else, are part of the real universe. That being the case, atomic physics should include more than just experimental results and their probabilities of occurrence. He professed that there are underlying atomic processes that do bring about the obtained result, and that causality is in effect on the atomic level, just as it is in a classical environment.

Yet, quantum theory doesn't concern itself with such inner details. It makes no reference to any underlying mechanism that brings about the experimental result. Although the experiment is the essential feature of quantum mechanics, the theory does not tell us how it works.

7.2   Einstein's Boxes

During his debate with Bohr, Einstein contrived many thought experiments that made evident his antipathy for entanglement and the non-locality that it engenders. His thought experiments were cleverly designed to show that quantum events could not possibly work in the bizarre way suggested by Bohr.

'Einstein's boxes' is one of those thought experiments. It questions the counterintuitive characteristics of entanglement by considering two boxes that are widely separated, but nonetheless, entangled. It is one thing to suggest that an electron and proton are entangled in the small confines of the hydrogen atom, but it is something else when one box is in Abilene and the other is in Boston.

Classically, the boxes experiment consists of one ball and two boxes, a preparation apparatus, and a measuring device that counts the number of balls in each box. During the preparation procedure in Chicago, the ball is randomly placed into one of the boxes and then one box (Box A) is mailed to Abilene and the other (Box B) to Boston. The preparation procedure is such that it is impossible to know where the ball is placed.

We want to determine the location of the ball. Because there is only one ball and, since balls do not multiply or disappear, the results obtained in Abilene are correlated with what we find in Boston: The single ball is a conserved entity, so that an empty box is always accompanied by a not empty box.

For example, if we find the ball in Boston, then the ball is obviously not in Abilene. We always find the ball in only one of the boxes.

Quantum mechanics and classical physics agree on these results. There is no controversy over what we will find when one of the boxes is opened. The disagreement is over how the ball ends up in one box and not in the other.

Classically, the preparation procedure does not favor one box over the other. 50% of the time the ball is placed in Box A and 50% of the time it is put into Box B. Unfortunately, we do not know which box gets the ball. If we did know, then we could do without probabilities. We see, again, that the need for probabilities is the result of our ignorance.

As soon as the ball is placed into Box B in Chicago, it is certain that the ball will be found in Boston. It never gets to Abilene. This result is predetermined by the action taken in Chicago. Although the human observer doesn't know where the ball is, rest assured, the ball is in Box B. Our ignorance has no effect on the whereabouts of the ball. This result is confirmed by opening a box after it reaches its destination.

When we find the ball in Boston, we know immediately that the Abilene box is empty. Such long distance correlations are common in our everyday experience because the results are known from the very beginning. Here, 'empty' and 'not empty' always occur together, all the way from Chicago to Boston and to Abilene. No one is bothered by the fact that the status of each box is determined at the moment one of them is opened.

Likewise, if the ball is placed into Box A, then the ball will be found in Abilene and the Boston box will be empty when it gets to Boston.

The status of each box is determined by opening only one of them. It doesn't matter which one we choose to open or how far apart they might be. We can learn about a distant box without interacting with it because we are observing a conserved entity and because the actual location of the ball was predetermined back in Chicago.

The classical physicist is at ease with all this. Separability and locality are not threatened by such long range correlations. Besides, the situation is easily understood: When we find the ball in Boston, then we know the ball was placed into Box B during the preparation procedure back in Chicago. There is nothing unusual about putting a ball in a box and the ball staying there until someone finds it at a later time. It is true that we don't know where the ball is until one of the boxes is opened, but where ever it is found, it was there all along.

Our ignorance of the ball's location before opening a box has nothing to do with the fact that the ball was definitely in the same box all the way from Chicago. It doesn't even matter whether or not anyone ever opens a box and looks inside. Everyone agrees that if the ball was put into Box B in Chicago then it remained there all the way from Chicago to Boston. Likewise, Box A was empty during its entire trip to Abilene. Just common sense, all this.

In the classical world with causality in effect, things happen for a reason. The reason why the ball was found in Boston is because some one in Chicago put it into Box B. It's as simple as that!

But the quantum explanation in terms of entanglement is not as straight forward, to say the least. If this box experiment obeys the rules of quantum mechanics, then something radically different is going on.

Quantum mechanics doesn't tell us anything about a ball being put into a box in Chicago. There is no mention of boxes moving from Chicago to Boston and to Abilene. Quantum mechanics describes only the situation when a box is opened. When that occurs, there are two possible results:

Result 1   Box A empty, Box B not empty.
Result 2   Box A not empty, Box B empty.

These are entangled results. Result 1, 'Box A empty, Box B not empty' is a single result. Box A and Box B are not two separate boxes. There is no 'Box A empty' and there is no 'Box B not empty'. There is only the entangled Result 1. Likewise, Result 2 cannot be separated into 'Box A not empty' and 'Box B empty'.

Quantum mechanics gives no reason why we get Result 1 or why we get Result 2. All we know is that an 'empty box' is always paired with a 'not empty box'. This correlation is now a consequence of quantum entanglement.

Something happened back in Chicago when the boxes were together that entangled the ball and both boxes into a single quantum entity, and the entanglement remained in effect all the way to Boston and to Abilene

Further, quantum theory denies that anyone put the ball into one of the boxes. Rather, it became part of the entangled whole. As part of the entangled apparatus, the ball and boxes no longer exist in space-time. Their individual realities are lost. We must continually remind ourselves that we do not describe what is happening to the individual parts of the apparatus. In spite of the great distance between them, the boxes are nonetheless a single entity.

The fact is, we should not be talking about a ball and boxes at all, because there is no ball and there are no boxes in the quantum description. At least, not the ordinary kind of balls and boxes we are familiar with. Rather the boxes are now an entangled pair, entangled with the ball and everything else. We should concern ourselves with that entangled whole. Quantum theory does not tell us how the ball ended up in the Boston box. It only tells us there is a 50% chance of finding it in Boston and a 50% chance of finding it in Abilene when one of the boxes is opened. Anything beyond this is pure speculation.

Contrary to how we do things in the real world, the quantum description is not concerned with an ordinary ball and two ordinary boxes. In the entangled wholeness those individual things no longer exist. Yet, many still ask, "Where is the ball before a box is opened?" There is no answer because in the quantum world the only ball is the ball observed when a box is opened. In this quantum event, there is no 'ball in the box' until we open a box and actually see 'a ball in the box'. There is nothing to describe until a box is opened.

Finding the ball in Boston, was not predetermined by what happened back in Chicago. Finding the ball in Boston, along with an empty box in Abilene, is a purely random event that happens without reason when a box is opened.

Whatever is going on here, it is obvious that the common sense approach inherent in classical physics is seriously violated.

In our real world description we had some prior knowledge about what happened in Chicago. But, the quantum theory denies this. Quantum theory makes no mention of any ball being put into a box in Chicago and then transported to Boston, while an empty box is sent to Abilene. There is no sequence of events that ends with a ball in Box B when one of them is opened. Quantum mechanics describes only the situation when a box is opened. It does not provide any space-time description of individual boxes moving to and fro.

In the absence of any pre-existing conditions, the only explanation is a voodoo doll kind of action-at-a-distance, so strongly opposed by Einstein. It is as if there is no box in Abilene until someone opens a box in far away Boston. At that instant an empty box jumps into existence in Abilene. You need not be a physicist to reject such a voodoo-like suggestion.

Then, how do we explain such a thing? We don't! There is no explanation. You should know by now that not knowing how an experiment works is part of the quantum way of doing physics.

Many are unhappy with this, but there are no inner details to tell you about, and we are wasting our time trying to answer such a question.

The quantum description recognizes only the results of the entangled experimental whole when a box is opened. All we know is that when a box is opened, we immediately get Result 1 or we get Result 2. The status of both boxes is known immediately. And, repeating the experiment many times, we get each possible result 50% of the time

Opening one of the boxes does not refer to only the box opened. Rather, it refers to the entangled whole. Opening the Boston box is equivalent to opening the box in Abilene. This is not what we imagine 'opening a box' to mean, but, remember again, we are not opening an ordinary box located in Boston that is separate from and independent of an ordinary box that happens to be in Abilene. Rather, we are actually observing the entangled pair. Opening either box gives us a measurement result for the entangled pair.

But the discussion gets harder to fathom when we consider that the boxes are spatially separated by thousands of miles. In the classical description that separation is real. It is the long distance involved that exasperates the discussion. Einstein believed that making something happen instantaneously in Abilene by opening a box in Boston violates his special theory of relativity and most physicists have a strong distaste for such voodoo-like behavior. Not only Einstein, but many physicists define 'nonlocal' to mean faster-than-light-speed communication. They refuse to accept quantum entanglement if it means the renunciation of relativistic locality.

But such a dilemma exists only if we insist that there are two separate boxes, classical things, one in Abilene and the other in Boston. Such a spatial separation is a classical concept that does not exist in the quantum description. It doesn't matter how great the spatial separation, the boxes are nevertheless inseparable!

In classical physics, spatial separation of objects is inherent in the space-time description of objective reality. But, in the quantum description there is no space-time description, and, therefore, no spatial separation and no threat to relativity theory. Realists need not worry.

Quantum mechanics does not describe two separate boxes moving cross country from Chicago to Boston and to Abilene. That would be a classical picture of what we imagine to be happening.

Nevertheless, Einstein had serious doubts about this quantum explanation. For him, the spatial separation is real and any non-local kind of communication was an action-at-a-distance prohibited by the special theory of relativity. Nonlocal events like this aren't supposed to happen. At least, not in the objective world that we are familiar with.

Quantum entanglement does enable non-local events, but without any faster-than-light communication because quantum events do not occur in the space-time of relativistic physics. The spatial separation between Boston and Abilene is a false representation of the quantum situation. It is an illusion based on what we see in our everyday experience. The separation between objects is a classical concept associated with the space-time description of real things, which does not apply to entangled quantum boxes.

Further, there is no physical interaction between the entangled boxes. There is no 'action-at-a-distance' in the relativistic sense. Actually, there is no 'action' and there is no 'distance' either.

Apparently, entanglement does not require any physical connection between the two boxes. They are not 'stuck together' by forces in the conventional sense.

In essence, this is nothing more than an extension of the Schrödinger's cat experiment to include two objects instead of one. And, once again, everyone agrees on the results, but we have no clear cut way to choose between the classical and quantum viewpoints.

And, as it has been all along, Einstein embraces the reasonable, common sense approach while Bohr talks of entanglement and all the strange, counterintuitive effects that go along with it.

7.3 Overview of the EPR experiment

By 1935 most physicists had accepted that quantum mechanics was inherently statistical. But most bothersome to Einstein was the fact that separability and locality, pillars of classical theory, had no place in the quantum description. Apparently, local realism was not required for the quantum physicist to describe atomic events.

It doesn't matter what you think of the moon. It is still there, moving as mandated by Newton's laws. Even if you never heard of Newton and spent your entire life underground, the moon would still be there, right where Newton says it should be. Why not with atoms? For Einstein, it made sense for individual atoms to be part of the objective universe.

The most important tenet of classical physics is that the physical universe is real and that everything in it obeys the physical laws. It is this objective reality that classical physicists successfully describe with their equations. As Einstein said, "Reality is the business of physics." He believed that all of science rests on this truth.

Yet, Bohr and his quantum mechanics deny any such thing.

Quantum mechanics is not about reality. It is not a space-time description. Because they are entangled, quantum things do not have a separate, real existence in space-time, as ordinary things do. Instead, they exist only when entangled with the measuring apparatus, and only when we are observing them!

The quantum particles being observed, as well as the measured values obtained in the experiment, are no longer independent of their measuring devices.

Bohr's complementarity principle provided the conceptual framework for the 'orthodox' interpretation of quantum theory, while Einstein became the leader of the opposition. He refused to accept the non-locality and non-separability in Bohr's interpretation. He set out to show that such a theory was necessarily incomplete because it ignored certain realities of atomic systems.

In 1935, Einstein, along with colleagues Boris Podolsky and Nathan Rosen at Princeton University, presented the greatest intellectual challenge ever to the Bohr interpretation. The arguments presented in their paper, commonly referred to as EPR, have caused more anguish among physicists, philosophers, and ordinary people alike, than any other in the history of science. Learned people still argue over it today.

So, does the moon really exist when no one is looking? The only true way to answer the question is to prove that the moon does, or does not, exist when we are not looking at it. We have to show that the existence of the moon is truly independent of any observers.

That is what Einstein, Podolsky, and Rosen aimed to do in an ingenious experiment in which we could 'see' on the quantum level without 'looking'. EPR would show that the value of a given attribute can be determined without actually measuring it. They would disprove the quantum axioms, "the only atom is an observed atom" and "the only value is the measured value". EPR intended to show those precepts to be false.

The basic idea behind the EPR experiment is to use a conservation law to provide a correlation between two widely separated particles. For EPR the spatial separation cannot be ignored. It means the particles are real.

They imagined an experiment involving two particles produced in such a way that the total momentum is conserved. This allows EPR to measure the momentum of only one of particles and to immediately know the momentum value of the far away second particle without 'looking at it'. Such correlations due to conservation laws are commonplace in real world experiments. There is nothing unusual about them.

But, they went further. Since the relative position of the particles is also conserved, they would use this same approach to determine the position value of the second particle. In so doing, EPR argued that a particle has definite values of both position and momentum in violation of the Heisenberg Uncertainty Principle.

Einstein and his colleagues did not say that quantum mechanics was wrong. They believed that if we are only interested in the experimental results, then quantum mechanics works just fine. But, it ignores everything else. And they believed that all that 'everything else' should be included in the quantum description. They would show that there is more to an experiment than just the results.

As an example, Einstein once asked, "Can we suppose that the decay of an unstable nucleus occurs at a definite moment in time"? It seems to make sense that if nuclear decay is a real event, then there has to be a definite moment at which the decay occurs. In fact, the decay time can be measured in a real experiment. A time-of-flight measurement of the decay radiation will give the instant at which the decay took place.

But, quantum mechanics only gives us the probability that a given nucleus will decay within a given period of time. We can calculate the half life of a radioactive substance, but we cannot predict the exact instant when an individual decay occurs.

Nowhere in the theory is there an exact instant at which a nucleus decays. Einstein argued, then, that the quantum description of nuclear decay is incomplete because it does not include the decay time.

Quantum theory says that nuclear decay is a random event and the decay time is indeterminate. There are an infinite number of possible decay times. An atom can decay quickly or not decay at all in a given time period. And there is no reason why one nucleus will decay and another does not.

There is no answer to the question, "When will the nucleus decay?", because there is no definite decay time. Further, there is no known process that causes nuclear decay. If we are to use quantum mechanics to describe atomic nuclei we will have to do it without such classical concepts.

In their experiment EPR intended to show that both position and momentum are real entities. A complete description of this experiment, then, would necessarily include both position and momentum, which quantum mechanics does not. Thus, they affirm that quantum mechanics is an incomplete description of reality and its statistical character is a result of that incompleteness. A complete quantum theory would be deterministic just as classical physics is. Position and momentum would have definite values.

7.4 The EPR experiment according to EPR

EPR intended to show that both position and momentum are 'elements of reality'. In classical theory, any physical property that can be predicted or measured with certainty is an element of reality. And in his Como lecture, published in 1928, Bohr said that classical phenomena "may be observed without disturbing them appreciably."

EPR agreed. In fact, they defined 'element of reality' thus:

> *If, without in any way disturbing a system, we can predict with certainty (i.e., with probability equal to unity) the value of a physical quantity, then there exists an element of reality corresponding to that quantity.*

This is often referred to as the EPR 'Criterion of Reality'. Here, the 'system' is a particle. This definition is a consequence of the separability principle, which requires that such elements of reality have definite values separate from and independent of the measuring apparatus. For EPR the 'value' of a physical attribute does not necessarily mean 'measured value' as asserted in quantum theory.

For example, the moon's momentum is an element of reality. The moon always has momentum as it orbits the earth, with or without being observed. You need not have a measured value in order for the momentum value to exist. I have no idea what the value of the moon's momentum might be. But, rest assured, it has a value nonetheless.

The crucial point is this – In classical physics, an element of reality exists with a definite value, known with certainty, without having to measure it. Since measurement requires an interaction, the trick is to find a way to determine its value, "without in any way disturbing" the particle. The EPR experiment was designed to do exactly that.

EPR further assert that in a complete theory,

> *every element of the physical reality must have a counterpart in the physical theory.*

They contend that if quantum theory omits an element of reality, then it is incomplete. Different elements of reality, being classical things, each have zero uncertainty, and there is no corresponding uncertainty relation between them. Yet, if they are non-commuting observables, at least one of them is excluded from the quantum mechanics description. Only one of them can have a definite value. Position and momentum are two such non-commuting observables. In quantum mechanics non-commuting observables are complementary attributes.

If the position is to be an element of reality, then it must have a definite value. The experiment done must be a position eigenexperiment and the describing quantum wave function is a position eigenfunction. Likewise, the momentum is an element of reality when we do the momentum eigenexperiment and its wave function is then an eigenfunction of the momentum operator.

But, the position eigenexperiment and the momentum eigenexperiment are mutually exclusive; it is impossible for the same experimental arrangement to be an eigenexperiment of both position and momentum.

This presents no difficulty in classical physics. We simply do separate measurements. The results of each experiment can then be combined to give us the position and momentum of the particle being observed. In classical physics we can use the results from mutually exclusive experiments in this way.

EPR is an imaginary experiment in which position and momentum are alleged to each have real values, rendering quantum mechanics incomplete while violating Heisenberg's Uncertainty Principle. Thus, for EPR, only one, not both, of the following statements is true:

    A. Quantum theory is incomplete.

    B. Heisenberg's Uncertainty Principle is valid.

It follows, then, that if quantum mechanics is complete, as alleged by Bohr and his followers, then the uncertainty principle is satisfied.

But, if EPR could show that the uncertainty principle is violated, then postulate B is false and postulate A is true; they will have shown that quantum mechanics is incomplete. A sketch of the EPR experiment is shown in Fig. 7.1. Alice and her experimental apparatus are on one side of a lab and Bob is on the other side. The particles are produced with equal momenta in opposite directions, possibly from a particle at rest that explodes into two equal masses. One piece moves towards Alice and her apparatus while the other piece moves in the opposite direction towards Bob on the other side of the lab.

Assuming no interaction of any kind after emission, each particle moves freely with constant momentum on its way to its detector. The two pieces are correlated because their relative position and their total momentum are conserved quantities. We know how this experiment works.

For Einstein, Podolsky, and Rosen, being classical physicists, all this was obvious. EPR tacitly assumed local realism, with separability and locality in effect, in their analysis. Alice and Bob are each doing their own experiments independently of each other. All the different experimental parts are real. In particular, each particle has its own identity and its own location in space-time. There really is a particle called 'A' and there is a second particle called 'B' just as shown in Figure 7.1.

The spatial separation is self-evident. We can see it in the lab. For Einstein, separability is a sufficient condition for Alice and Bob to each have an independent reality that enables us to describe Alice's experiment without any reference to what Bob is doing. Alice and Bob can each make individual measurements at the same time or at different times. It doesn't matter.

Locality, on the other hand, implies that any action taken by Alice has no effect on Bob's results. Remember, there is no connection between the two. Separability means that Alice and Bob exist on their own, while locality ensures that there is no 'spooky action at a distance' going on between them.

In hindsight, we recognize that EPR was actually an attack on Bohr's non-separability principle, what we now call entanglement. Entanglement is the antithesis of separability and locality. EPR hoped to avoid entanglement of the two spatially separated particles with their careful definition of 'elements of physical reality'.

THE EPR EXPERIMENT

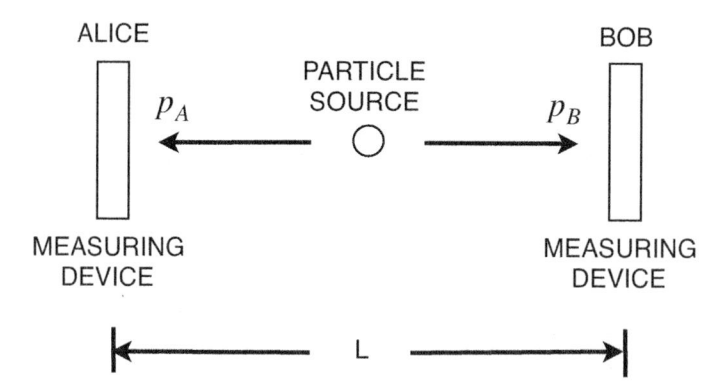

Figure 7.1. Here, p_A is the momentum of Alice's particle and p_B is the momentum of Bob's particle. Alice measures the momentum of Particle A. Conservation of momentum gives $p_B = -p_A$ and because the distance between detectors is fixed, $x_B = L + x_A$ when Alice measures the position. Thus, Alice knows the values of position and momentum belonging to Particle B without in any way interacting with it. Bob's position and momentum are, then, elements of reality as defined by EPR.

The particles are produced at the source with zero total momentum, $p_A + p_B = 0$, where p_A is the momentum value of Alice's particle and p_B is the momentum value of Bob's particle obtained in Bob's experiment. Particle A had the value p_A at every instant as it traversed its way through the apparatus to Alice's detector. When Alice gets the momentum value p_A, she is measuring a value that was there all along.

Likewise, Bob's momentum $p_B = -p_A$, from conservation of momentum, had that value all the way from the source to Bob's detector. It doesn't matter whether he measures it or not. In fact, it is not necessary for Bob to do anything. He could be on his lunch break or forget to turn on his measuring device. Because of conservation of momentum, we do not need to measure Bob's momentum directly in order to know its value. Conservation of momentum makes it all possible. Alice can determine the momentum of Bob's particle "without in any way disturbing it". Bob's momentum is then, by definition, an 'element of reality'. If Bob does measures his particle's momentum we are certain that he will get the value $p_B = -p_A$. Nothing unusual here.

Alice then reconfigures her apparatus to measure the position. She always gets the same value x_A and immediately knows the position of Bob's particle to be $x_B = L + x_A$. Again, all measurements made by Alice are with her apparatus acting only on her particle on her side of the laboratory. She has determined the position of Bob's particle, without doing anything to it. Therefore, the position of Bob's particle is also an 'element of reality', along with its momentum.

Locality ensures that Alice's position measurement will have no effect on Bob's momentum value. So, Bob still has the momentum value $p_B = -p_A$. Conservation of energy is still in effect. Bob's particle is not affected in any way by any changes made by Alice. Once emitted, it always has the value $p_B = -p_A$. It doesn't matter what Alice is doing. There can be no effect on Bob's particle as a result of a non-interaction between Alice and Bob. Obviously, a non-interaction cannot change anything.

An applied force is needed to change momentum. And, here, there isn't any. If its momentum has not been changed, then Bob's particle must have had the same momentum value since inception, long before Alice did anything. This is the only reasonable explanation.

Furthermore, EPR assert that the Heisenberg uncertainty principle fails because both position and momentum have definite values. It follows that quantum mechanics is incomplete because it ignores momentum when we measure the position, even though we know the momentum value from an earlier experiment. Likewise, it ignores position when the momentum value is known. According to EPR, both position and momentum would be included in a complete theory.

EPR claim that, as a result of Alice's measurements, Bob's particle has definite values for both position and momentum, just like other real objects. Heisenberg's uncertainty principle fails and quantum mechanics is incomplete!

7.5 The EPR experiment according to Bohr

Leon Rosenfeld was with Bohr in Copenhagen when they received a copy of the EPR paper. He remembered that it "came down upon us as a bolt from the blue." All other work at Copenhagen came to a halt while Bohr set out to undo what he considered to be a grave misunderstanding of quantum theory. He scrutinized the EPR experiment in fine detail and then reformulated it in his own terms. It took him six weeks. When finished, he commented to Rosenfeld, "They do it smartly, but what counts is to do it right."

Bohr then gives what he considers to be the correct interpretation of the EPR experiment in terms of complementarity and entanglement. He argued that the customary view of causality and physical reality is now compromised by quantum entanglement. At the time, however, the term 'entanglement' was not known to him. Schrödinger would not coin that term for another year.

Bohr describes Alice and Bob as part of a single entangled entity. Alice and Bob, both their apparatuses, both particles, and both results are all entangled into a single experimental whole, in spite of the spatial separations seen in the lab. Our sensory perceptions fool us into believing they are all separate from each other and functioning in the usual way.

Entanglement denies the existence in space-time of all the individual parts, including the particles. In the entangled description, there is no space between Particle A and Particle B. Actually, there is no space-time, no Particle A and, no Particle B. There is only the pair of particles entangled along with everything else.

Bohr argued that the EPR definition of element of reality is flawed. As a consequence of entanglement and the Heisenberg Uncertainty Principle, there is an essential ambiguity in the expression, "----without in any way disturbing a system---".

Bohr agreed that Alice can determine values for Bob's position and momentum by making measurements on her particle without interacting with Bob in any way. Nevertheless, that does not mean that we know both position and momentum simultaneously. Bob's momentum value is known only when Alice performs a momentum eigenvalue experiment. Otherwise it is not known to anyone. Likewise, Bob's position has a known value only when Alice measures the position.

For Bohr, the momentum value obtained by Alice in her momentum eigenexperiment refers to the entangled pair. There is never a value of Alice's momentum alone without a correlated value for Bob's momentum.

For Bohr, $p_A, p_B = -p_A$ is a single result of the momentum eigenexperiment. These values are inseparable. They always appear together when momentum is measured. You cannot have one without the other and it does not matter who does the actual measurement. This correlation is a manifestation of entanglement.

Classically, $p_B = -p_A$ only when momentum is conserved. This requires a definite momentum value. So, momentum is conserved only when Alice does the momentum eigenexperiment. When she does the position eigenexperiment, the momentum is extremely uncertain. Alice does not know her own particle's momentum, let alone knowing Bob's. It doesn't matter that she had made her position measurement "---- without in any way disturbing" Bob's particle. Any prior knowledge of the momentum is meaningless. Remember Peres! The only momentum value is the measured value, which requires a momentum eigenexperiment. There is no momentum value associated with the position eigenexperiment.

EPR was wrong to assume that momentum conservation was still in effect when Alice measured the position.

In the position eigenexperiment, Alice's particle momentum has no value. Consequently, conservation of momentum is lost and Bob's momentum is completely unknown. Likewise, when she does a momentum eigenexperiment, we have lost any knowledge of her position value because her position eigenexperiment is now an unperformed experiment. So, we do not know where the particles are when $p_B = -p_A$.

Bohr says that in the EPR experiment we can measure position or we can measure momentum, but not both at the same time. They are complementary attributes. Alice's position eigenexperiment generates the entangled position values $x_A, x_B = L + x_A$ and her momentum eigenexperiment yields the entangled momentum values $p_A, p_B = -p_A$.

But we cannot combine these results to get a well defined momentum $p_B = -p_A$ at well defined position $x_B = L + x_A$ as claimed by EPR. There is no answer to the question,"What is Particle B's momentum at position x_B?"

EPR did acknowledge that their position and momentum measurements were mutually exclusive. But, according to Einstein, this shouldn't matter. If we always obtain the same value for Bob's momentum every time we measure it, they argued that Bob's momentum must always have that value, even when we are not measuring it. That obtained momentum value should not be affected by Alice's position measurement. Locality forbids it. And, after all, Alice did do all of this "without in any way disturbing" Bob's particle.

In classical physics it is not unusual to use data from mutually exclusive experiments to obtain a more complete picture of the system in question. For example, since the moon has a trajectory, we have no qualms about ascribing definite values of both momentum and position to the moon. No reasonable person would question this.

Further, argued EPR, Alice has a choice of which eigenexperiment to perform. According to the Heisenberg Uncertainty Principle, she can measure momentum and render Bob's position value to be completely uncertain. On the other hand, if she chooses to measure the position, then Bob's position does become an element of reality with a definite value and zero uncertainty.

So, according to quantum theory, as understood by Einstein, Podolsky, and Rosen, Bob's position is either real with zero uncertainty, or it has an indeterminate value with infinite uncertainty depending on what Alice decides to do with her apparatus. And in either case, she does absolutely nothing to Bob's particle! It is one thing for Alice's actions to effect the properties of her own particle, the one that she is interacting with, but, it is something else entirely for her to instantly impose a reality, or a non-reality, as the case might be, on Bob's far away particle without ever interacting with it. Voodoo physics, indeed! For Einstein, Podolsky, and Rosen, this was utter nonsense. They conclude that,

> *No reasonable definition of reality could be expected to permit this.*

This is true, of course, for any "reasonable definition of reality". But quantum mechanics does not pretend to define reality. This statement is true for classical physics, but it does not apply to quantum mechanics. Yet, Einstein believed otherwise. For him, quantum mechanics, like all of science, is supposed to describe an objective reality.

In our new quantum way of thinking, there is no separate Alice and no separate Bob. There is no disturbance between them. As in other things quantum, they are part of an entangled Alice-Bob experimental whole. EPR was wrong to think of them as real separate entities.

Bohr does admit that Alice's particle does have a peculiar 'influence' on Bob's particle, but not in the conventional sense, not in the manner of interacting particles. Rather, Bohr would come to discuss that 'influence' in terms of a 'wholeness' that we now call 'entanglement'. Bohr's alternative explanation in terms of quantum entanglement, allows correlations between widely separated particles in a way that violates both separability and locality. For Einstein, this was unacceptable.

As we have seen elsewhere, Einstein argues for a reasonable and straight forward explanation of what is going on, while Bohr attacks our sensibilities and challenges our preconceived notions about physical reality. Nevertheless, most physicists at the time believed that Bohr "did do it right".

7.6 The Bohm version of the EPR experiment

Most modern discussions of the EPR experiment refer to the version first proposed by David Bohm. It is often referred to as EPRB. The EPRB experiment is done with spin 1/2 particles where Alice and Bob each use Stern-Gerlach magnets to make spin 1/2 measurements. The z-component and the x-component of the spin vector, which are non-commuting observables, take the place of position and momentum in the original EPR experiment. The analysis is a bit easier because there are only two possible spin values, +1 and −1.

In this version of the EPR experiment, identical spin 1/2 particles are produced at the source with zero total spin, $S_A + S_B = 0$, where S_A is the spin of the particle in Alice's detector and S_B is the spin of Bob's particle. Spin conservation always gives $S_B = -S_A$, provided that Alice and Bob have their magnets oriented in the same direction.

We initially assume both magnets are oriented in the z-direction. There are two possible results:

Result 1. $S_A = +1, S_B = -1$

Result 2. $S_A = -1, S_B = +1$

When Alice gets Result 1, she immediately knows that the z-component of Bob's spin is, $S_B = -1$. This correlation is due to spin conservation, which allows Alice to know the value of Bob's spin without "in any way disturbing" it. According to EPRB, Bob's z-component is an element of reality.

Alice and Bob now align their magnets in the x-direction, allowing them to measure the x-component of spin. They proceed to show that the x-component is also an element of reality. They conclude that both the x- and z-components have definite values. But the quantum description must omit one of them since they are non-commuting observables.

As in the original EPR experiment, EPRB conclude that the quantum description must be incomplete because both spin components have definite values in defiance of quantum theory.

So, who are we to believe? We have the classical description of the EPR experiment that embraces local realism, while the quantum explanation depends on entanglement and the Heisenberg Uncertainty Principle. Here again, Einstein and Bohr agree on the results, but they disagree on what is the correct interpretation.

The non-locality of long distance entanglement, first brought to our attention by the EPR paper, has bedeviled physicists ever since. We have an overwhelming number of classical experiments done over the past one hundred years that confirm that nonlocal interactions do not occur in nature, at least not in the world we are familiar with. Consequently, physicists do not believe in voodoo dolls!

And yet, everyone agrees that quantum mechanics does predict that entangled systems can behave in non-local ways even when separated by vast distances. But the question still remains: "Does nature really behave in such a non-local way on the atomic level?" In 1935 we were not yet able to do the experiments that would answer the question for us.

Surprisingly, after 1935, Einstein rarely mentioned 'elements of reality' in his opposition to quantum mechanics. And, although probabilities are not intrinsic to Einstein's universe, it was not the deciding factor in his lifelong objection to quantum theory. Even though his comment about, "god playing dice", has been widely cited in the literature, he did come to accept the statistical nature of quantum events.

The more serious issue for Einstein, was quantum mechanics' rejection of separability and locality. He never did accept the loss of objective reality. He always maintained that the independent reality of spatially separated things was essential to our objective description of nature. He expected that a realistic description of the atomic world would eventually be found.

8

BELL'S THEOREM
The second revolution begins

Realism is dead, -----, for realism is well and truly dead.
                                                Arthur Fine

It is a pity that Einstein's idea doesn't work. The reasonable thing just doesn't work.
                                                John Stewart Bell

8.1 Introduction to Bell's theorem

The original EPR experiment did not challenge the statistical nature of quantum mechanics. Its only issues with quantum mechanics were the questions of reality and completeness. Einstein believed that an objective reality is essential to physics, while quantum mechanics has no such requirement.

In their conclusion EPR professed that a complete quantum theory of physical reality is possible, but they did not offer any suggestions on how to accomplish it.

Einstein never wavered in his belief that quantum things are real, maybe extremely small, and even unimaginable, but real nonetheless. He believed that atoms and atomic processes are part of the known objective reality. Atomic events should occur in space-time and obey the principles of separability and locality, as does everything else in the physical universe.

Physicists had to wait thirty years after EPR for the conclusive work of John Stewart Bell before they could make a definitive comparison between a reality based quantum description and the conventional theory based on entanglement. Most importantly, his modified EPRB experiment was doable in a real world laboratory.

His realistic version of quantum mechanics embraced local realism with separability and locality in place, just as was done in the original EPR experiment. In so doing, he hoped to make the quantum experiment real while generating the same statistical results, which were never in question.

If successful, quantum probabilities would be a consequence of not knowing the inner workings, just as it is with classical probabilities. Quantum experiments would then have a realistic underpinning that, although hidden from us, would behave in a deterministic way. Causality would then be inherent in the real processes underlying quantum theory.

But, contrary to what he hoped for, he proved just the opposite. Bell's theorem states that there are no hidden inner workings of the quantum experiment and a reality based quantum mechanics is unable to replicate the predictions of a conventional quantum calculation. Quantum mechanics does not embrace the local realism of classical physics, and, most importantly, we cannot make it so. A classical explanation for the statistical nature of quantum events is impossible. It is futile to think otherwise.

But Bell did provide us with a test known as 'Bell's Inequality', which enables us to distinguish between a reality version of quantum mechanics and its entangled counterpart. Here, we will derive a simple form of this inequality in terms of probabilities.

The literature is full of thought experiments used to obtain Bell-like inequalities. The basic approach in them all is to construct an experiment that is a composite of real experiments. Then assume that the so constructed composite experiment is also real. In such a case, everything is separable and there is no entanglement.

But in conventional quantum mechanics the combined experiment is often an experimental whole, an entangled thing. It is no longer real and the principles of separability and locality do not apply. Rather, the original real experiments are lost in the entangled compound experiment. So, we are not surprised when the results of classical EPR-like experiments agree with Bell's inequality while the results of entangled experiments do not.

Yet, Bell's approach seems so logical and so simple, that many insist that nature cannot possibly behave otherwise. Unfortunately, actual real world EPR-like experiments done over the past forty years show that nature does indeed 'behave otherwise'.

The example used here to derive Bell's Inequality does not involve any physics. We simply read two data sheets individually and then read them together. The individual data sheets maintain their individual identities when read together. We don't expect any strange goings on in such a straightforward realistic endeavor.

The results don't prove or disapprove any theorems. We will later use this same approach on Bell's modified EPRB experiment.

8.2 A Bell-like Experiment - comparing true-false test results

Let us consider the following common place situation: Alice and her friend Bob have taken a certain true-false exam that contains twenty questions. As with all such tests, there are only two possible answers to each question - True (T) or False (F). This is comparable to a two-state observable that upon measurement yields one of two possible results.

When the scores are released, they talk and discover that each has the same score. Each had two incorrect answers. They do not know which specific questions they got wrong, only that each was wrong on two out of the twenty questions.

We do not have copies of their individual answer sheets. We only know that there is a 10% chance, two out of twenty, that any one of Alice's answers is incorrect. Thus, the probability that any one of Alice's answers is wrong is

$P_A(\text{wrong}) = \frac{2}{20} = .1$. Likewise, for Bob, $P_B(\text{wrong}) = \frac{2}{20} = .1$.

We emphasize that $P_A(\text{wrong}) = .1$ belongs to Alice's answer sheet alone. It has nothing to do with Bob. Alice's test results are free and independent of whatever Bob did on his test. In fact, it doesn't matter whether Alice talks to Bob or not. Even if Bob never takes the test, Alice still has $P_A(\text{wrong}) = .1$. Anything Bob does to his answer sheet has no effect on Alice's test results.

Likewise, $P_B(\text{wrong}) = .1$ has nothing to do with Alice and her test score. Bob's test results are his own doing, independent of whatever Alice has done.

All this is a consequence of separability and locality. Nothing unusual here.

Imagine, now, that we examine the two answer sheets together. Can we predict how many questions will have incorrect answers on the combined sheets? That is all we want to know. We simply look for incorrect answers on Alice's and Bob's answer sheets taken together.

We will do this by inspection alone. No advanced math required! But, to do this correctly, since we haven't seen the actual answer sheets, we need to compare all possible pairs of answer sheets that have two wrong answers each. But, this is not as tedious as it sounds.

The combined answer sheets might look like Possible Answer Sheets 1 shown in Table 8.1. This is only one possibility. We have no way of knowing if the individual sheets shown correspond to the actual answer sheets belonging to Alice (A) and to Bob (B). Most likely, they do not. But they each have two wrong answers, which is our only requirement.

POSSIBLE ANSWER SHEETS 1

Q	1	2	3	4	5	6	7	8	9	10	11	12	13	14	15	16	17	18	19	20
A	T	T	F	F	**T**	T	F	T	T	F	F	F	**F**	T	F	T	T	F	T	F
B	T	T	F	F	F	T	F	T	**F**	F	F	F	T	T	F	T	T	**T**	T	F
A B					•				•				•					•		

Table 8.1. The first row is the question number. The second (A) row is Alice's answer sheet and the third row (B) is Bob's. Wrong answers are shown in large bold type. Alice has wrong answers to questions 5 and 13. Bob had questions 9 and 18 wrong. The combined sheets (AB) have wrong answers, shown as dark circles, to questions 5, 9, 13, and 18.

It does not matter if we examine the answer sheets individually or together. The recorded data remains the same. The individual answer sheets do not change in any way when we read them simultaneously or one at a time. Alice has wrong answers for questions 5 and 13 in Row A and also, in the combined Row A and Row B. Likewise, Bob's wrong answers are always at questions 9 and 18.

So, why do we bother mentioning something so obvious? It is such a trivial observation that, normally, it wouldn't warrant our attention. We do mention it because some combined quantum things are entangled with a loss of individual identities. Individual characteristics no longer exist in the entangled whole. If this was a quantum experiment, the individual answer sheets would be lost in the entangled whole. Rows A and B in Table 8.1 would no longer exist.

We saw this with entangled routes in our discussion of interference in Chapter 6. But, in this example, the individual answer sheets are obviously not entangled. Answer sheets and combined answer sheets are not quantum things. The individual sheets, are still separate classical (real) things when read together. Separability is the norm and they are still separate, even when combined. Alice still has wrong answers for questions 5 and 13 in the combined sheets. Bob's wrong answers still appear at 9 and 18. Nothing is changed from the original individual sheets, as expected in the real world.

There are four questions with wrong answers in the combined sheets. Alice and Bob each contribute two wrong answers to the total of four.

But Possible Answer Sheets 1 is not the only possibility. For example, Alice and Bob could have each given the wrong answer to question #5, as shown in Possible Answer Sheets 2 .

POSSIBLE ANSWER SHEETS 2

Q	1	2	3	4	5	6	7	8	9	10	11	12	13	14	15	16	17	18	19	20
A	T	T	F	F	**T**	T	F	T	T	F	F	F	**F**	T	F	T	T	F	T	F
B	T	T	F	F	**T**	T	F	T	T	F	F	F	T	T	F	T	T	**T**	T	F
A B					•								•					•		

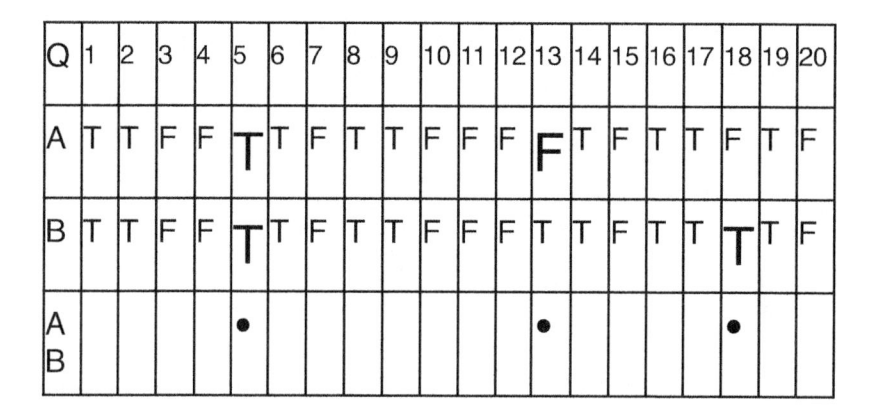

Table 8.2. Another possible set of answer sheets for Alice and Bob, each with two wrong answers (large bold type). But, here, they both get question 5 wrong. Their combined wrong answers appear on only three questions, 5, 13, and 18.

Under these circumstances, with one common wrong answer, it is possible to have combined sheets with only three questions with wrong answers.

There is only one other possibility. Alice and Bob could have identical answer sheets, as shown in Possible Answer Sheets 3 where they both get questions 5 and 13 wrong. In that case, it is possible to have only the two questions with wrong answers on the combined sheets.

These are the only possible scenarios. So, how many questions with a wrong answer do we find when we inspect both of their answer sheets together? There is no definite answer. There are three possibilities. We can have two, three, or four questions with wrong answers on the combined sheets.

POSSIBLE ANSWER SHEETS 3

Q	1	2	3	4	5	6	7	8	9	10	11	12	13	14	15	16	17	18	19	20
A	T	T	F	F	T	T	F	T	T	F	F	F	F	T	F	T	T	F	T	F
B	T	T	F	F	T	T	F	T	T	F	F	F	F	T	F	T	T	F	T	F
A					•								•							
B																				

Table 8.3. Here, Alice (A) and Bob (B) have identical answer sheets. They both get questions 5 and 13 wrong. When we examine the sheets together, all the wrong answers appear on only those two questions, 5 and 13.

In terms of probabilities,

$$P_{AB}(\text{wrong}) = \frac{2}{20}, \frac{3}{20}, \text{ or } \frac{4}{20}.$$

Here $P_{AB}(\text{wrong})$ is the probability that any question in the combined sheets will have a wrong answer. That's it! That is all we wanted to know.

In any case, it is never possible to have more than four questions with wrong answers on the combined sheets, the two from Alice and the two from Bob. This makes sense because all the wrong answers had to originate with either Alice or Bob.

This can be written

$$P_{AB}(\text{wrong}) \leq \frac{4}{20} = .2.$$

This is a Bell inequality for this example. You can make up any answer sheets you like for Alice and for Bob, provided that each has two wrong answers, and you will always find four or fewer questions with wrong answers when you examine them together. Common sense all this.

There is a way to check Bell's theorem for this example. We can obtain true copies of the answer sheets for Alice and Bob and then compare them. We are confident that the true copies read together will have four or fewer questions with wrong answers. It cannot be any other way. Bell's inequality will be satisfied. Guaranteed!

In terms of the individual probabilities, Bell's inequality is written

$$P_{AB} \leq P_A + P_B.$$

This is the general form that we will use in what follows.

Bell's inequality is a reality test. If separability still exists in the combined experiment, then P_A and P_B will apply to the combined experiment and the Bell inequality will be satisfied. Nothing unusual here.

8.3 Bell's experimental test for local realism

We now apply this same approach to Bell's modified Bohm version of the EPR experiment. In the original experiment Alice and Bob had their Stern-Gerlach magnets set in the same direction. But, in Bell's version, Alice and Bob have their magnets aligned in different directions.

Since they are now measuring different spin components, spin conservation is no longer in effect and it is now possible for Alice and Bob to get same-spins, + + or − −, as well as opposite spins + − or − +.

However, the opposite-spin values are now a consequence of entanglement.

In the entangled description of Bell's altered EPRB experiment, there are four possible results of a spin measurement corresponding to four entangled eigenvalues:

Result 1. $S_A = +1, S_B = +1$
Result 2. $S_A = +1, S_B = -1$
Result 3. $S_A = -1, S_B = +1$
Result 4. $S_A = -1, S_B = -1$

Same-spin values occur in Results 1 and 4. Opposite values in Results 2 and 3.

With Alice's magnet set at angle θ_A and Bob's magnet at θ_B, quantum mechanics gives the probabilities

$$P_{AB}(\text{same-spin}) = \sin^2\left(\frac{\theta_{AB}}{2}\right)$$

$$P_{AB}(\text{opposite-spin}) = \cos^2\left(\frac{\theta_{AB}}{2}\right),$$

where $\theta_{AB} = \theta_B - \theta_A$. Note that it is θ_{AB}, not the individual angles, θ_A and θ_B, that determines the probabilities. Only the relative angle, θ_{AB}, has any significance.

We first calculate the probabilities for the original EPRB, a classical experiment where Alice and Bob are doing separate Stern-Gerlach experiments with $\theta_A = 0°$ and $\theta_B = 0°$. Here, spin conservation is in effect and they never get same-spins.

This configuration is shown in Figure 8.1 and possible results are shown in Table 8.4 where classical separability is evident. Alice's distribution of spin measurements in the first row is separate and independent from Bob's data in the second row. We see, by inspection, that spin conservation gives $P_{AB}(\text{same-spin}) = 0$.

THE EPRB EXPERIMENT

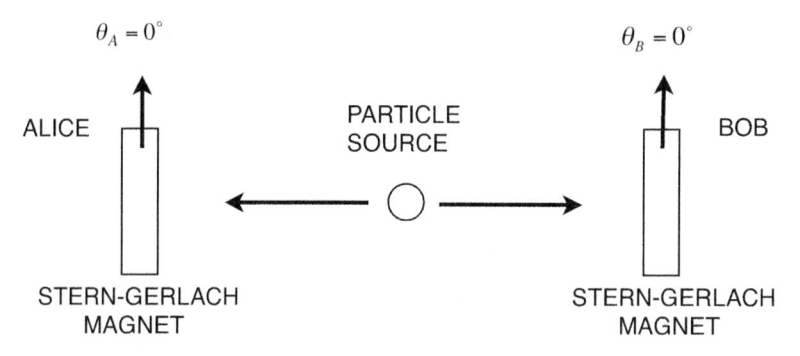

Figure 8.1. This is the Bohm version of the classical EPR experiment with the detectors not shown. Alice and Bob each have their magnets oriented in the same direction so that $\theta_{AB} = \theta_B - \theta_A = 0°$. For this experimental configuration, spin conservation gives $S_A + S_B = 0$. Alice and Bob never get same-spin values.

Bob makes his measurements. Alice does hers, unaffected by what he might be doing. Each has his/her own distribution of spin values. If Alice has shut down her experiment, Bob's data remain the same.

But, in Bohr's description of EPRB, Alice and Bob and all the apparatuses on both sides of the lab, including the two data sheets, are part of an entangled whole. There is no pictorial representation of the entangled EPRB experiment and Figure 8.1 does not apply. The individual features of all the parts, including the data sheets are lost. There are no longer any individual data sheets to be shown in a table analogous to Table 8.4, which shows data from a classical experiment.

In the entangled EPRB experiment only the quantum probabilities have any meaning.

THE EPRB EXPERIMENT RESULTS

ALICE $\theta_A = 0°$	+	-	+	+	-	-	+	+	-	-	+	-	+	-	-	+	-	-	+	+
BOB $\theta_B = 0°$	-	+	-	-	+	+	-	-	+	+	-	+	-	+	+	-	+	+	-	-
SAME SPINS							N	E	V	E	R									

Table 8.4. Some possible results obtained in the Bohm version of the EPR experiment. This is a classical experiment with both magnets aligned in the same direction. Alice and Bob accumulate data randomly, but with $\theta_A = \theta_B = 0°$, they always get opposite-spin values in the same measurement. This correlation is due to spin conservation.

In the entangled EPRB experiment, with $\theta_{AB} = 0°$, the quantum probabilities are

$$P_{AB}(\text{same-spin}) = \sin^2(0°) = 0$$

$$P_{AB}(\text{opposite-spin}) = \cos^2(0°) = 1.$$

Quantum mechanics predicts that, as a consequence of entanglement, Alice and Bob always get opposite spins in the original EPRB experiment. It doesn't matter if we assume the EPRB experiment to be real or entangled. The classical and quantum descriptions both predict the same results. The real experiment with separability and spin conservation in effect replicates the quantum results with entanglement. This was the result that Bell had hoped for. But, unfortunately, $\theta_{AB} = 0°$ is a special case.

With $\theta_A = 0°$ and $\theta_B = 0°$, Alice and Bob are never found with same-spin values. Obviously, neither Alice nor Bob are responsible for any same-spin values, since there are none.

Bell takes the classical approach where Alice is real and separate from Bob and where locality demands that Alice's results depend on her magnet setting alone. When $\theta_A = 0°$, she always gets the same results, the ones shown in Table 8.4 and, further, she does not contribute to any same-spin values.

Likewise, when $\theta_B = 0°$, real Bob gets the same results shown in Table 8.4 and he is not responsible for any same-spins.

Bell recognized that same-spin values are possible only if $\theta_{AB} \neq 0°$, which requires a change in the orientation of at least one of the Stern-Gerlach magnets.

We proceed just as we did above with the analysis of test scores. But, rather than observing wrong answers on test sheets, we now look for same-spin values that result when $\theta_{AB} \neq 0°$. Specifically, we consider a Bell version of the EPRB experiment with

$$\theta_{AB} = \theta_B - \theta_A = 60° - (-60°) = 120°.$$

But, we treat Alice's experiment with $\theta_A = -60°$ separately from Bob's experiment with $\theta_B = 60°$, as is done in classical physics.

We first determine P_A(same-spin), the quantum probability that Alice alone generates a same-spin value when $\theta_A = -60°$. This requires that $\theta_B = 0°$, as shown in Figure 8.2.

REAL BELL EXPERIMENT A

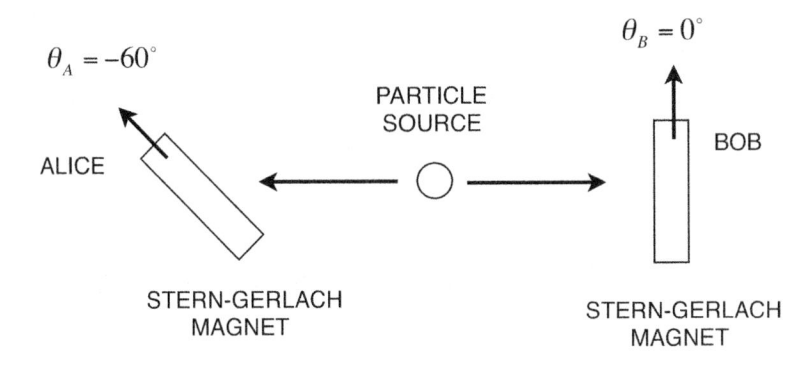

Figure 8.2. Alice has set her magnet at $\theta_A = -60°$, while Bob leaves his magnet unchanged at $\theta_B = 0°$. In this configuration, Alice and Bob measure different components so the total spin is no longer conserved, making it possible for Alice and Bob to get same-spins (+ +) or (- -) in some measurements. All same-spin values are due to $\theta_A = -60°$. Bob does not contribute to same-spins with $\theta_B = 0°$. This is a classical experiment with locality and separability in effect.

Thus,
$$P_A(\text{same-spin}) = P_{AB}(\text{same-spin}) \text{ when } \theta_B = 0°.$$
We get
$$P_A(\text{same-spin}) = \sin^2\left(\frac{0° - (-60°)}{2}\right) = .25.$$

Some possible results are shown in Table 8.5. Comparing Alice's data with Bob's we count five same-spin values, as expected. Each same-spin value occurs wherever there is a change in Alice's data.

Alice contributes five same-spin values out of 20 measurements when $\theta_A = -60°$ and $\theta_B = 0°$. In what follows, whenever $\theta_A = -60°$, locality demands that Alice always gets the distribution of spin values given in Table 8.5.

REAL BELL EXPERIMENT A RESULTS

ALICE $\theta_A = -60°$	+	-	+	-	-	-	+	+	**+**	-	+	-	-	-	-	+	**+**	**+**	+	+
BOB $\theta_B = 0°$	-	+	-	-	+	+	-	-	+	+	-	+	-	+	+	-	+	+	-	-
SAME SPINS				•				•				•					•	•		

Table 8.5. The results of Real Bell Experiment A, where Alice has reset her magnet to $\theta_A = -60°$ resulting in the five changes shown in large bold type. Spin is no longer conserved and same-spin values appear where ever the changes occur in Alice's data. Bob's results have not changed from $\theta_B = 0°$. He does not contribute to any same-spin values. We simply compare Alice'a data sheet, the first row, to Bob's, given in the second row, to see where same-spins occur.

We then calculate the quantum probability for getting same-spins in the experiment where

$$\theta_{AB} = \theta_B - \theta_A = 60° - 0° = 60°.$$

This is Real Bell Experiment B shown in Figure 8.3. Here, any same-spin values are a consequence of changes in Bob's data.

REAL BELL EXPERIMENT B

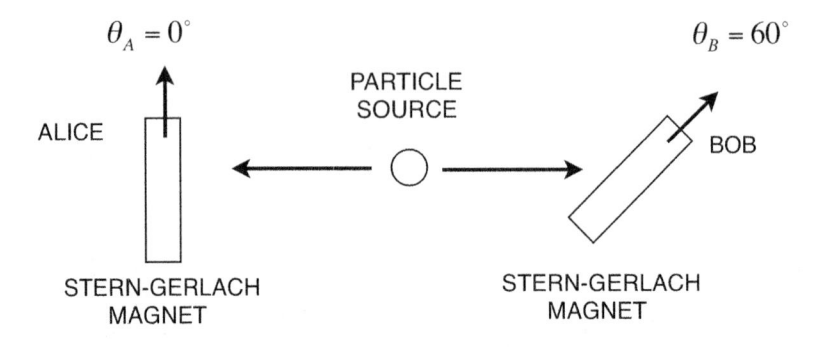

Figure 8.3 Here, Alice has reset her magnet back to $\theta_A = 0°$, her no-same-spin configuration, while Bob has changed his magnet setting to $\theta_B = 60°$. The total spin is no longer conserved and any same-spin values are attributed to changes in Bob's data.

Proceeding as we did in the analysis of Real Bell Experiment A, the quantum probability of getting a change in Bob's data with his magnet at $\theta_B = 60°$ is

$$P_B(\text{same-spin}) = \sin^2\left(\frac{60° - 0°}{2}\right) = .25.$$

A possible distribution of spin values for Real Bell Experiment B is shown in Table 8.6. Comparing Alice's data with Bob's we count five same-spin values, as predicted. Same-spins occur where ever a change occurs in Bob's data.

REAL BELL EXPERIMENT B RESULTS

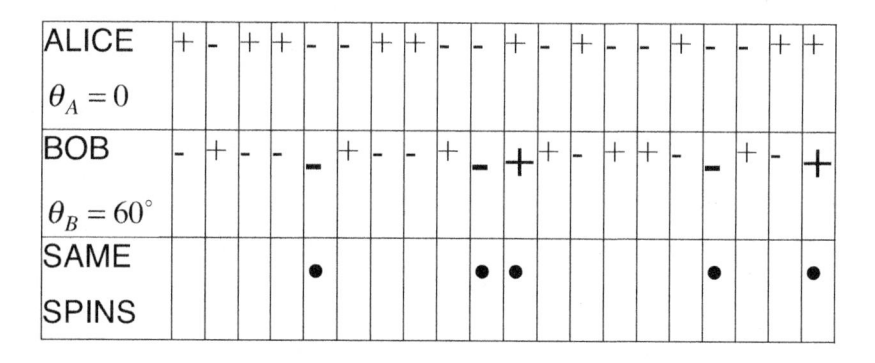

ALICE $\theta_A = 0$	+	-	+	+	-	-	+	+	-	-	+	-	+	-	-	+	-	-	+	+
BOB $\theta_B = 60°$	-	+	-	-	**−**	+	-	-	+	**−**	**+**	+	-	+	+	-	**−**	+	-	**+**
SAME SPINS				•					•	•						•				•

Table 8.6. Bob's magnet at $\theta_B = 60°$ generates five changes from his $\theta_B = 0°$ no-same-spin data. Those changes are shown in bolder large font size. For each change there is a corresponding same-spin value. Assuming local realism is in effect, Alice with her magnet at $\theta_A = 0°$ does not contribute to any same-spins.

There is, again, a 25% chance of getting same-spins in any given spin measurement whenever $\theta_A = 0°$ and $\theta_B = 60°$. We attribute these five same-spin values to changes in Bob's data.

We now construct Real Bell Experiment AB with $\theta_A = -60°$ and $\theta_B = 60°$. This experimental configuration is shown in Figure 8.4.

Although Real Bell Experiments A, B, and AB are mutually exclusive, we use the results from Real Bell Experiments A and B in our analysis of Real Bell Experiment AB, as is routinely done in classical physics.

REAL BELL EXPERIMENT AB

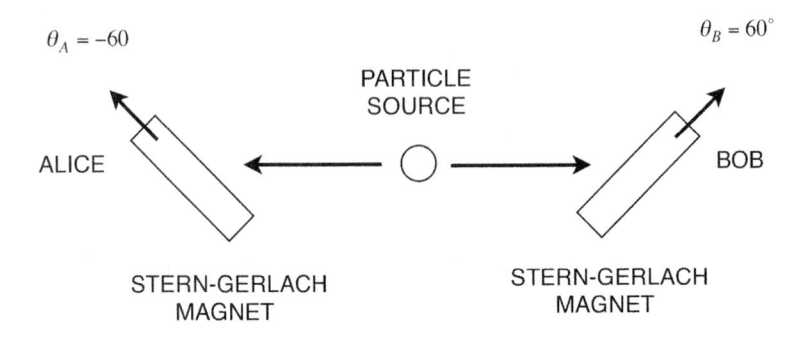

Figure 8.4. Real Bell Experiment AB consists of Alice's experiment on the left and Bob's experiment on the right. This is a classical experiment where Alice and Bob each contribute separately to the production of same-spins.

We now apply Bell's inequality to Real Bell Experiment AB:

$$P_{AB}(\text{same-spin}) \leq P_A(\text{same-spin}) + P_B(\text{same-spin})$$
$$\leq .25 + .25 = .5.$$

Bell's inequality maintains that there will be ten or fewer same-spins in Real Bell Experiment AB.

We expect that Alice will have five changes from her no-same-spin results whenever her magnet is set at $\theta_A = -60°$. Likewise, Bob always has five changes with $\theta_B = 60°$. This is shown in Table 8.7.

REAL BELL EXPERIMENT AB RESULTS

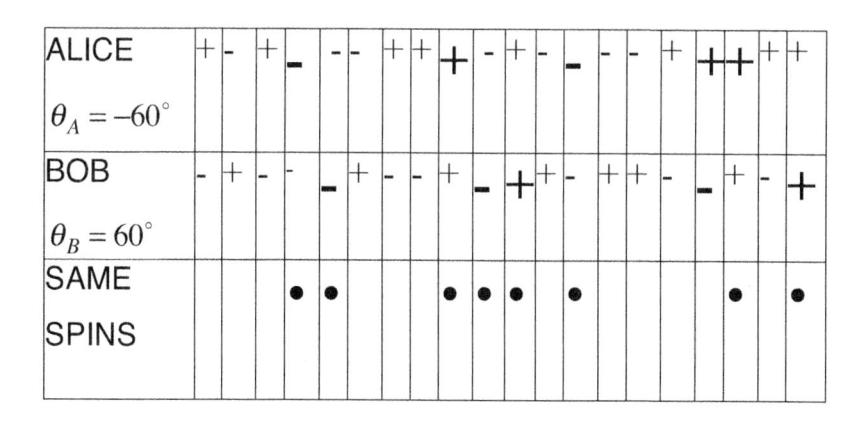

ALICE $\theta_A = -60°$	+	-	+	-	-	-	+	+	+	-	+	-	-	-	-	+	+	+	+	+
BOB $\theta_B = 60°$	-	+	-	-	-	+	-	-	+	-	+	+	-	+	+	-	-	+	-	+
SAME SPINS				•	•		•	•	•		•						•			•

Table 8.7 This is the data expected if Alice and Bob are doing separate experiments, Alice with $\theta_A = -60°$ and Bob with $\theta_B = 60°$. Alice's data was obtained from Real Bell Experiment A, Bob's data from Real Bell Experiment B. Comparing their data sheets shows eight same-spin values in agreement with Bells inequality, which predicts no more than ten same-spin values.

We count eight same-spins corresponding to a same-spin probability of $P_{AB}(\text{same-spin}) = \frac{8}{20} = .4$. Bell's inequality is satisfied, as it should be for real experiments.

We have determined how many measurements yield same-spins by looking at the data and counting. That is all there is to it! Classically, there is no reason not to do it this way. How can we go wrong by simply looking and counting? The results of Real Bell Experiment AB satisfy the Bell inequality.

But we have not yet calculated the quantum probability for getting same-spin values in Bell Experiment AB. The hope was that the quantum calculation would also satisfy Bell's inequality. This would imply that we can construct real EPR-like experiments that mimic the entangled versions.

8.4 The quantum calculation with entanglement

In the quantum description, Bell Experiment AB with $\theta_{AB} = \theta_B - \theta_A = 120°$ is a single entangled whole. We no longer treat Alice's experiment and Bob's experiment as real separate entities. There are no individual experiments in the quantum description of Bell Experiment AB.

Bell had hoped that a conventional quantum calculation for same-spin probabilities with $\theta_{AB} = 120°$ would also satisfy his inequality. This would demonstrate that quantum mechanics can be made locally realistic without changing its predicted probabilities, which were never in doubt. Then, EPR-like experiments will have passed the reality test! Bell would have successfully introduced objective reality into quantum mechanics, which would now embrace the classical tenets of separability and locality.

We could then describe how the quantum experiment works in classical terms. Alice, her apparatus, Particle A, Bob, his apparatus, and Particle B, would all be real separate things, part of the objective reality that makes up the classical universe.

This would mean that a quantum experiment does have inner workings that are ignored by conventional quantum mechanics, which would then be deemed incomplete as envisioned by Einstein, Podolsky, and Rosen. Atoms would be real things. Einstein would win the reality debate.

But, before we declare a winner, we must do the quantum calculation for the entangled Bell Experiment AB. With entanglement in effect, there are no results from now unperformed Real Bell Experiment A and there are no results from unperformed Real Bell Experiment B. Simply comparing results is no longer an option, since we have no data from those earlier experiments to compare.

In the quantum description there is no P_A(same-spin) from Real Bell Experiment A and no P_B(same-spin) from Real Bell Experiment B that can be used when analyzing Bell Experiment AB. When we do Bell Experiment AB, those probabilities are meaningless numbers inferred from unperformed experiments.

Alice's data obtained when $\theta_A = -60°$ and Bob's data obtained when $\theta_B = 60°$ no longer apply when $\theta_{AB} = 120°$. There is no data to be recorded in a table analogous to Table 8.7.

The ultimate question to be answered is, "Does the quantum probability for same-spins satisfy Bell's inequality?" If so, then the reality based Bell Experiment AB is equivalent to the entangled version. Our assumption that Alice and Bob each individually contributes to the creation of same-spin values would be justified. There would then be a classical description of the quantum experiment.

There would be no need for entanglement and its inherent non-locality. This is the result that Einstein would have liked.

Bell's inequality assumes that Alice's results depend only on the settings of her apparatus. Likewise, for Bob. Locality demands it, and Bell wanted to replicate the quantum predictions with locality in effect.

But, in his 1964 paper, despite having hidden variables in place, Bell recognized that Alice's results depend on the relative angle $\theta_{AB} = \theta_B - \theta_A$, not on θ_A alone. Her results do depend on θ_B, the setting of Bob's magnet, in violation of the locality principle. Bell could not avoid the non-locality associated with quantum events.

In the entangled version of Bell Experiment AB the relative angle between magnets is $\theta_{AB} = 120°$ and the probability of getting same spins is

$$P_{AB}(\text{same-spin}) = \sin^2\left(\frac{120°}{2}\right) = .75.$$

This purely quantum calculation, with entanglement in effect, predicts that Alice and Bob do get same-spins 75% of the time, while Bell's inequality indicates that Alice and Bob can never generate same-spins more than 50% of the time. Bell's inequality is violated! The quantum Bell Experiment AB fails the reality test. The quantum description cannot be made to include local realism.

The entangled Alice and Bob expect same-spin values 15 times out of 20 measurements. But, we had seen in earlier experiments that separate Alice and Bob can contribute a maximum of only ten. Where do those extra five same-spin values come from?

Again, there is no answer to such a question. It is meaningless. This is not classical physics and there doesn't have to be a reason why we get an extra five same-spins. It just is! All we can say is that in the quantum description of Bell Experiment AB with $\theta_{AB} = 120°$ we expect 15 same-spin values out of 20 measurements. There is no further explanation.

Bell's approach does not yield the quantum predictions as he wished. Rather, he unintentionally proved that

quantum mechanics is incompatible with local realism.

Quantum mechanics truly is different from classical physics. It is not about reality and we cannot make it so.

Bell's theorem states that there can be no underlying reality in the quantum version of Bell Experiment AB. If there was, then Bell's inequality would be satisfied. But it is not.

Bell had hoped that a hidden reality could be added to quantum mechanics, making it deterministic, no different from classical physics. There would be an underlying reality that obeyed the classical laws. Bohr, on the other hand, took the extreme view that there is no inner reality, no individual parts. We can never know how quantum experiments work because there are no inner details to be known.

It was the grand wish of many, not only Bell and Einstein, that we could make quantum mechanics complete by injecting such an underlying reality into the quantum description. Then, quantum 'weirdness' would be explained away by the every day physics of Newton, Maxwell, and Einstein. But, it was not to be.

And, most disturbing to many, including Bell, is that quantum mechanics is fundamentally non-local. As Bohr suggested in his response to EPR, entangled Alice does have an instantaneous effect on her far away partner. In the quantum world, entanglement is the decisive factor and reality is lost!

We must accept that 'spooky action-at-a-distance', so abhorred by Einstein, is an essential characteristic of quantum events. Many find this conclusion to be appalling. Einstein himself said, "My physical instincts bristle at that suggestion."

Many different versions of Bell's inequality have been obtained in more rigorous ways than what was done here. But the results are always the same: The quantum calculations violate the Bell inequality, regardless of its form.

8.5 The empirical evidence

Although Bell failed to replicate the quantum predictions, he did provide us with a way to determine which description of nature is correct. We can now do real world versions of Bell Experiment AB, where we simply record the data and determine the probabilities without any reference to reality or entanglement. This will demonstrate that nature actually agrees with the realistic approach professed by Einstein and Bell or that it agrees with the entanglement view of Bohr.

Thought experiments like EPR have a long history of challenging and prodding physicists into unknown areas. But, physics is, at its core, an empirical science and all questions must ultimately be answered in the laboratory. So it was with Bohr and Einstein with their differing interpretations of the EPR experiment. The outcome of the reality vs. entanglement debate would be settled by actually doing Bell Experiment AB in the laboratory.

In 1972 John Clauser and Stuart Freedman did just that. They performed the first experimental test of Bell's inequality. They performed Bell Experiment AB with photons in entangled polarization states. Photon polarization is a two-state observable analogous to spin 1/2.

Their measured probability violated Bell's inequality. It did appear as if nature obeyed the rules of quantum mechanics.

Other early experiments also favored quantum mechanics, with few exceptions. But most were plagued by low detector efficiencies. In some cases, too many photons were not being counted, and, at other times, there were counts with no photons in the detectors. There was also the nagging skepticism over the non-locality inherent in the quantum description. We had to be sure that there was not some sort of mysterious force between Alice and Bob that allowed one to influence the other's results. We needed an experiment in which such a process is impossible.

Bell had shared this concern. In the conclusion of his 1964 paper, he stated that any definitive test should be a delayed choice experiment where Alice changes her magnet setting so rapidly, at the last possible moment, that it would be impossible for any real communication to occur between her and Bob. We had to be sure that there was no signaling between them. Here, 'real communication' means signaling at speeds less than the speed of light, as mandated by the special theory of relativity. If the quantum predictions are still affirmed in such a 'no signaling' experiment, then there would be no saving the principle of locality.

In 1982 Alain Aspect and colleagues in Orsay performed a series of experiments still considered to be the definitive tests of Bell's theorem. All known loopholes, including the no-signaling requirement, were painstakingly eliminated. Their results agreed with the quantum predictions. This time the Bell inequality was surely violated!

After Aspect, it was generally accepted that local realism is not supported by the empirical evidence. Many later experiments have confirmed this conclusion.

In some modern versions of Bell Experiment AB, the spatial separation between Alice and Bob has been made intentionally extreme. No-signaling tests have actually been done with Alice and Bob in different towns. Alice resets her magnet at the last possible moment ensuring that no signal from Alice can reach Bob before he detects his particle. Again, the presence of quantum non-locality is confirmed.

We are forced to accept the unthinkable. Quantum entanglement and its inherent non-locality is a fact of nature and any attempt at a classical explanation is fruitless.

Bell Experiment AB does not consist of two separate experiments, one done by Alice and one done by Bob, in spite of what we see in the lab. Alice and Bob are entangled in a non-local way, and remain so, no matter how great the spatial separation between them.

What we see is not what we get! The spatial separation between Alice and Bob seen in the lab is an illusion. With entanglement in place, Alice and Bob are an inseparable part of the experimental whole.

But this does not mean that 'nonlocal' is synonymous with faster-than-light-speed signaling. Entanglement does not support such an idea. Signals sent between things require they be separated by some distance in space-time. But quantum things do not exist in space-time. They are not 'separated' and there is no 'distance' between them in the entangled whole. Thus, the quantum non-locality seen in long range correlations does not require signaling of any sort.

But, then, we must renounce any space-time description of quantum entities: *An existence in space-time is a classical concept that does not apply to quantum things.*

As much as we would like it to be otherwise, there is no objective reality on the atomic level. Atoms are not just tiny things that happen to exist in microscopic space-time. They do not obey the principles of separability and locality, and quantum events truly do occur for no reason.

Modern EPR-like experiments continue to close any possible remaining loopholes, and Bell's inequality continues to be violated. These experiments confirm that local realism, as professed in classical theory, is untenable. If we continue to insist otherwise, then our classical calculations will continue to fail us and we will be forever confronted with all kinds of imagined weird behavior.

Entanglement, not local realism, is the essential feature of the atomic world.

The continued violation of Bell's inequality in contemporary experiments indicates that we cannot scale down to the atomic level and expect things to work as usual. The realm of atoms and the like is not a miniature version of the macroscopic universe that we are familiar with. There is too much evidence to think otherwise. The common sense approach of Einstein, so successful in the real world we are familiar with, is simply wrong.

Even if we knew nothing about quantum mechanics, the results of these experiments would still violate Bell's inequality, effectively dispelling the classical concept of an objective reality forever. Whatever form atomic theory might take in the future, it will not embrace the local realism so strongly advocated by Albert Einstein.

The empirical evidence confirms that quantum mechanics works just fine without an objective reality. In its stead, we are left to ponder only results and probabilities.

So be it!

FURTHER READINGS

Selected Books

Niels Bohr, *The Philosophical Writings of Niels Bohr,* Volumes I, II, III and IV. Ox Bow Press Reprint (1987)

J. S. Bell, *Speakable and Unspeakable in Quantum Mechanics.* Cambridge University Press (2010)

Louisa Gilder, *The Age of Entanglement: When Quantum Physics was Reborn.* Vintage Books (2009)

Nick Herbert, *Quantum Reality.* Anchor Books (1985)

Arthur Fine, *The Shaky Game: Einstein Realism and the Quantum Theory.* The University of Chicago Press (1996)

Selected Articles

The 'Bohr-Einstein Dialogue' and the 'Commentaries' printed in John Archibald Wheeler and Wojciech Hubert Zurek, ed, *Quantum Theory and Measurement.*
Princeton University Press (1983)

Albert Einstein, *Quantum Mechanics and Reality.* Dialectica **2** 3-4, 320-324 (1948)

Christopher A. Fuchs and Asher Peres, *Quantum Theory Needs no Interpretation.* Physics Today **53**, 3, 70 (2007)

N. David Mermin, *Is the Moon There When Nobody Looks? Reality and the Quantum Theory.* Physics Today **38** 38 (1985)

J R Minkel, *Quantum Theory Fails Reality Checks.* Scientific American, April 18, 2007

ABOUT THE AUTHOR

Tom Marcella is the grandson of illiterate Italian immigrants. He was the first in his family to go to college. He received his PhD from Boston College and taught physics at UMass Lowell for thirty four years. Trained as a nuclear physicist, he worked with the accelerator group at UMass Lowell, but his main areas of interest became the second law of thermodynamics and the foundations of quantum mechanics. Other than family and physics he has enjoyed being around airplanes. He has flown them, helped to build one, and occasionally jumped from them. He says life has been great fun and he is thankful to all family, friends, colleagues, and especially his teachers who helped make it all possible. He presently lives with his wife in an in-law apartment in Massachusetts during the summer and in a retirement condo in Florida during the winter.

Printed in Great Britain
by Amazon